Computer Graphics
Systems and Applications

Managing Editor: J. L. Encarnação

Springer
Berlin
Heidelberg
New York
Barcelona
Budapest
Hong Kong
London
Milan
Paris
Santa Clara
Singapore
Tokyo

Jürgen Schönhut

Document Imaging
Computer Meets Press

With 88 Figures, 38 in Color

 Springer

Dr. Jürgen Schönhut
Fraunhofer-Institut
für Graphische Datenverarbeitung IGD
Rundeturmstraße 6
D-64283 Darmstadt

Library of Congress Cataloging-in-Publication Data

Schönhut, Jürgen.
    Document imaging : computer meets press / Jürgen Schönhut.
        p.   cm. -- (Computer graphics--systems and applications)
    Includes bibliographical references and index.
    ISBN 3-540-62045-1 (hardcover : alk. paper)
    1. Document imaging systems.   I. Title.   II. Series.
HF5737.S36   1997
006.4'2--dc21   97-26264
                                                                   CIP

ISBN 3-540-62045-1 Springer-Verlag Berlin Heidelberg New York

© Springer-Verlag Berlin Heidelberg 1997
Printed in Germany

Typesetting: Camera-ready by the author
Cover: Künkel + Lopka, Heidelberg
SPIN: 10519522   33/3142 - 5 4 3 2 1 0 – Gedruckt auf säurefreiem Papier

Acknowledgements

As with many projects in life this book would not have been possible without the help of many individuals. Special thanks are due to Prof. José L. Encarnação for his encouragement during the time of writing. The history of the creation of this book dates five years back. Starting with Microsoft Word (later for Windows) showed problems with typesetting of formulas; the advice of Springer-Verlag to produce the book in LaTeX turned out to show different problems – color images could not well be integrated. In the end we decided to use PageMaker, but it still was not an easy task.

Therefore I happily acknowledge the invaluable help of the following people: Jörg Zedler for the support going through several iterations of tools to produce this book including some of the more complex illustrations, Karin Zürner for producing many of the illustrations and helping with layout, Uwe Schneider with the attempt to put everything into LaTeX and with typesetting formulas, Haimo Fritz and Marwan Ramadan for producing some illustrations, Stefan Daun for support in various aspects, Michael Kokula for the idea how to illustrate certain details in image rastering, Cornelia Nestmann for putting colorimetric data in machine readable format, all the rest of my department for support and contribution of ideas and critics, Dr. Michael Macedonia for reading most of the text for correct English, and also for his helpful suggestions.

This book could also not have been produced without the patience of my wife and my children; they excused me in many occasions to give me the time to bring this book to its completion.

Darmstadt, May 1997

Table of Contents

1 Preface

This book was written while heading the department *Document Imaging* at the Fraunhofer Institute for Computer Graphics in Darmstadt/Germany. It developed under the pressure that introducing new personnel to the field is a rather time-consuming task. Most co-workers coming from the university had good background in one of the related areas; however, the communication across the fields always was difficult. The author expects that this task can be significantly simplified by using this book.

Document Imaging is a new discipline in Applied Computer Science. It is building bridges between Computer Graphics, the world of prepress and press, the areas of color vision and color reproduction. A book on Document Imaging needs a rather broad focus given the areas involved and related.

For readers with background in Computer Graphics this book gives insight into all problems related to putting information in print, a field only very weakly covered in textbooks of Computer Graphics.

This book is the first to give a comprehensive overview of Document Imaging, the areas involved, and how they relate.

2 Introduction

Document Imaging as used in this book deals with the process of mapping of electronic information to an output device; this includes all technologies involved with this process. An important component of Document Imaging is Raster Image Processing, a term used mostly in prepress context meaning processing some page to the point when it can be reproduced in a conventional printing process. With the advent of Electronic Publishing (EP) Raster Image Processors (RIPs) have become a key component in the electronic processing of documents for printing and viewing, and thus in Document Imaging. Since the introduction of PostScript more than a Decade ago, and after long discussions whether it can fulfil the quality requirements of professional printing, this Page Description Language (PDL) has become one of the most important components in RIPs of today. Therefore a large portion of this work will center discussion around the conceptual framework of PDLs, PostScript in particular, and show the integrative functionality of RIPs today.

2.1 A Simple Reference Model for Document Imaging

In order to localize the discussions on Document Imaging a simple framework shall help us to understand the pieces and the relationships between them.

This small framework shows the role of RIPs. RIPs obtain all information from the creative systems and transform it into printed pages. This process includes integration of different sources from creative systems as well as all the necessary adaptations and corrections for the final printing process. The following figure is an illustration of the processes associated with Document Imaging.

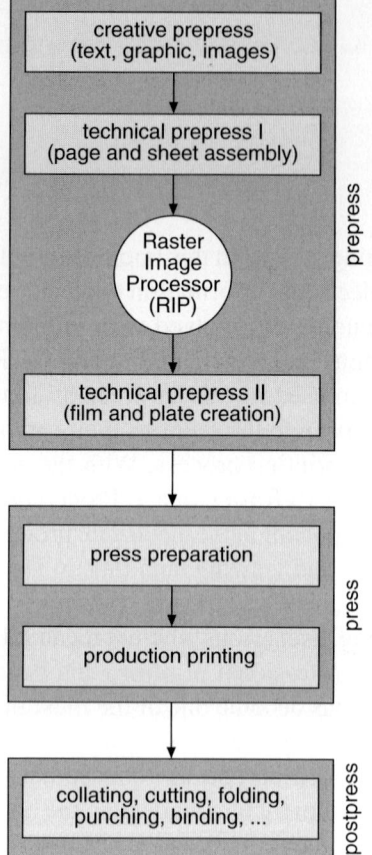

2.2 Topics Covered

This book will not discuss all the different applications for creative systems; the number of available software in this area is very large, including versions of the same software systems for different platforms. We will also exclude details of the postpress area from the scope covered. However we will cover aspects of press and their relation to the prepress world. Computer integrated manufacturing (CIM) has become a key to future developments in this area. We will discuss some of the issues to be solved in the near future to allow CIM to be applied to printing.

Further we will discuss in this book the formats coming into the RIP, their integration into one single output product, all the processes required to transform these formats into device pixels on film, plate, or direct imaging press devices, as well as on displays, and on computer printers. Among the formats coming into RIPs we will deal briefly with PostScript, TIFF, IIF, CGM, and some prepress exchange formats.

The transformation to the device includes the proper use of color. The chapter on color (chapter 3) contains a thorough overview of the phenomena related to our visual system, as well as the difficult problems related to color reproduction. An intensive discussion of halftoning methods is included. This discussion contains simple thresholding approaches as well as amplitude modulated and frequency modulated screening methods and derivatives. The effects of dot shapes are also discussed. The well-known approach of Neugebauer forms the basis to theoretical models for the printing process.

Chapter 4 covers the effects of the processes involved in offset printing and different aspects important to produce good quality. The principles of color and density measurement as practiced in relation to offset printing are discussed. The chapter centers around the whole production process for print products; it deals with all steps at different level of detail. Among other topics it summarizes the achievements in the area of RIPs. It discusses available solutions as well as research directions regarding output quality and speed. Conclusions are drawn from the topics already discussed, and future directions are pointed out. Among others the Open Prepress Interface (OPI) is discussed as an important piece in the production chain. Moving prepress data objects to press are all dealt with in some granularity. In particular problems related to computer integrated manufacturing of print products are the guiding path in this discussion, explaining how information available at the prepress stage can be made available at the press, and during finishing, and how to avoid redundant reacquisition of information at later stages.

A chapter on formats (chapter 5) is included to provide a better insight into today's Babylon of ways to store and exchange information. This contains a brief discussion of the most important and most promising formats available today.

The still remaining problem areas are summarized, the fields where there is still a lack of good solutions. Today's advertisements suggest that all problems have been solved. Chapter 6 shows that this is still an illusion.

Some chapters are dedicated to a very special problem which is present in every single application in prepress: the integration of information (chapter 7) obtained from different sources and applications. Besides the statement of the problem, example solutions are given to the integration problem that have been developed in research labs. A discussion dedicated to the problems of information interchange (chapter 8) for the production is included. It deals with the requirements for application independent information interchange and shows some solutions in practical examples.

The last chapter (chapter 9) deals in form of an outlook with questions related to multiple use of information for different media products. Cross media publishing is not the center of this book, however the questions also have influence on print production solutions.

A comprehensive bibliography at the end of this book provides access to other literature for further and more detailed study.

A glossary forms the basis to use this book as a source of reference.

3 Color Phenomena in Display and Printing

With the advent of so many different color output devices color has become a key issue both for display and printing. Given the brilliant colors available on display monitors the request for equivalent capability in printing has become more prevalent, but probably not achievable. Matching color across devices is a requirement that is not easy to fulfil. In order to understand the issues involved we will discuss in the following sections the phenomena related to color and help understand the problems and possible solutions.

3.1 Introduction

Color is the result of a variety of phenomena that are very common and almost always subject of very poor understanding. In most cases it is a leftover from our early school experiments with paint: Blue and Yellow mixed results in Green. Maybe later with color television we have heard about Red, Green and Blue as monitor colors. And of course we have heard about spectral colors, the colors of the rainbow from our physics lessons. The latter is probably the reason why most people believe color is a physical phenomenon.

To say it right at the beginning: color is a phenomenon of perception; physics, chemistry and physiology only help to build a model and give explanations to some aspects. Color is only color when it is perceived! Of course wavelengths of light play an important role here, but only the interaction of light

7

with our visual system makes a color to be a color (perception). Before we go into detail about color vision, we will look at some experiments showing the complexity of our system of visual perception.

*Ill. 3.1:
Experiment on
Simultaneous
Contrast*

Take a table which is divided in two halves vertically; one half is blue, the other half is red. In the center of each half is a neutral gray square. The gray square in the blue environment looks yellowish, the gray square in the red surrounding looks greenish. This is the case although both gray squares are colored with exactly the same color.

*Ill. 3.2:
Experiment on
Simultaneous
Contrast Black and
White*

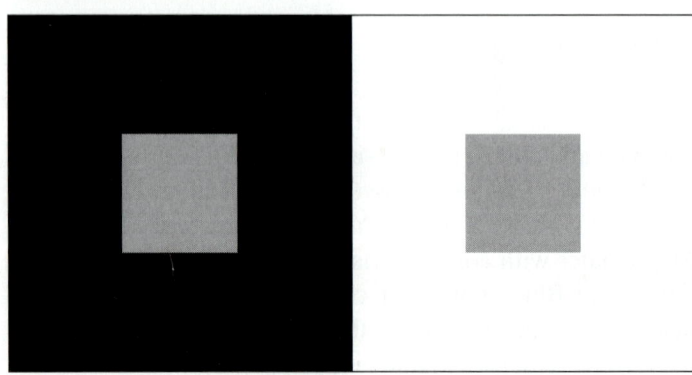

This phenomenon is called simultaneous contrast, showing that we perceive different colors while the inks on the surface and the lighting conditions are exactly equal.

This experiment also works with black and white surrounding; then the gray in the white surrounding is perceived as darker than the gray in the black surrounding. Altogether these experiments show something of the importance the environment has on the perception of a color.

Ill. 3.3:
Underwater
photograph taken
without flash

Ill. 3.4:
Underwater
photograph taken
with flash

A simple experiment in the colorful underwater world: If you ever tried to take underwater photographs without flash equipment you surely have been disappointed by the results: you saw a very colorful fish that you took a picture of, and the result looks more like a monochrome blue-green picture. If you had taken this photograph with a flash, the beautiful colors of the fish would have been clearly visible (if it were close enough). What has happened? Light travelling in water for a distance of more than five meters has no orange and red spectral components left (which the colors on the film document); the water absorbs these wavelengths and serves as a filter.

So objectively there is no light with wavelengths from the red end of the spectrum present; still we can perceive red colored objects. The reason for this is the white balance of our visual system; it categorizes the available light as being white (after some time of adjustment); this is also the reason for the need to calibrate video cameras to the correct white point. The color perception now is depending on the perceived white and shifts colors significantly for our perception. The examples 3.3 and 3.4 on page 9 show the effect.

3.2 The Human Eye

Basic to the understanding of color is the construction of the human eye. The picture 3.5 shows a schematic section through the human eye. The ray coming from real world objects are focused by the lens on the retina. The retina contains light receptive cells which are basic to the understanding of the phenomena.

*Ill. 3.5:
Cross section
through the
human eye*

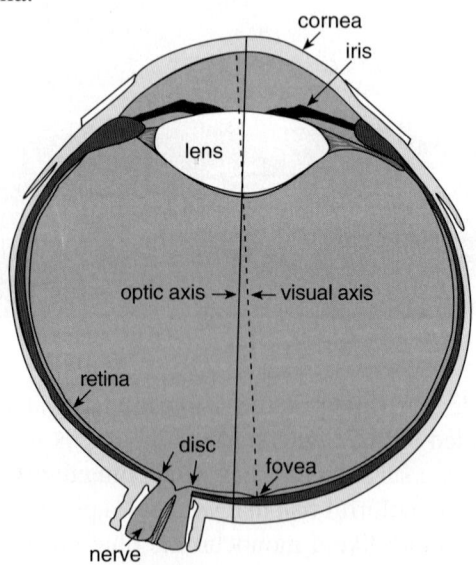

3.2.1 Retina: Rods for Brightness and Cones for Chromaticity

We distinguish four kinds of such light receptors, one called rods for bright/dark distinction and also for vision under extremely low lighting conditions, and three kinds of receptors with different sensitivity to light of different wavelengths, called cones.

The number of rods by far exceeds the number of cones, and rods and cones are not evenly distributed over the retina. Most of the cones are placed close to the optical center (fovea) of the eye, the rods are very densely spaced mostly in the peripheral regions of the retina. The figure 3.6 shows the distribution of rods and cones along a horizontal cut through the eye. For our understanding of color vision we are mainly looking at the cones. The specific wavelength sensitivity of the different types of cones is the foundation of color vision.

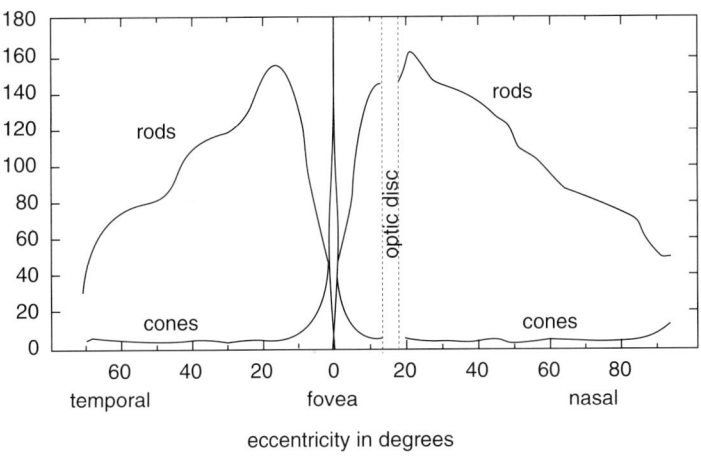

Ill. 3.6:
Distribution of
rods and cones
(in 1000/mm²)
in the retina along
a horizontal cut
through the eye

3.3 Eye Response Functions

Measurements show that the three kinds of cones have different absorption spectra with their maxima roughly at 445 nm, 535 nm, and 600 nm. These different absorption characteristics also lead to different response functions. Due to the varying distribution of cones on the retina different angles of view lead to different perceived colors. The *Commission Inter-*

nationale de l'Éclairage (CIE) therefore defined two color-imetric standard observers, the so-called 2°-observer (CIE 1931) for small angles of view of up to 4°, and the 10°-observer (CIE 1964) for wider angles of view above 4°. An object of 1.75 cm diameter seen at a viewing distance of 50 cm is seen under a 2° viewing angle, a 10° view angle is achieved with an object of 8.75 cm diameter respectively. The resulting response functions for equal energy spectrum colors of the three kinds of cones $\bar{x}$, $\bar{y}$, and $\bar{z}$ for a 2° standard observer are shown in the following diagrams 3.7–3.9.

Ill. 3.7:
Eye response
functions $\bar{x}$ for
a 2° observer

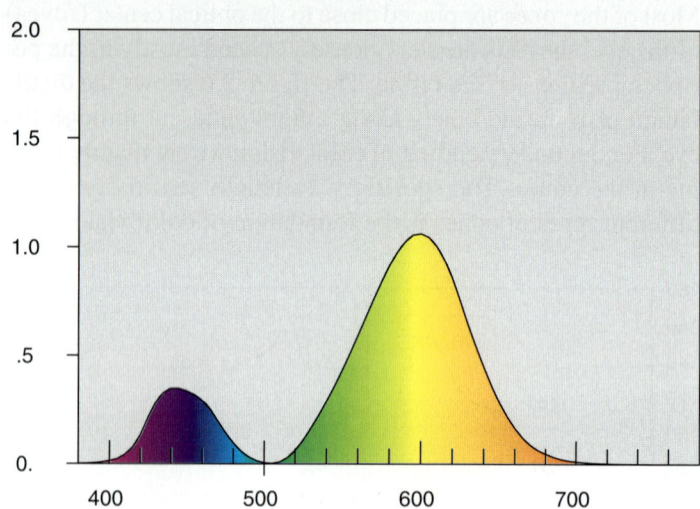

Ill. 3.8:
Eye response
functions $\bar{y}$ for
a 2° observer

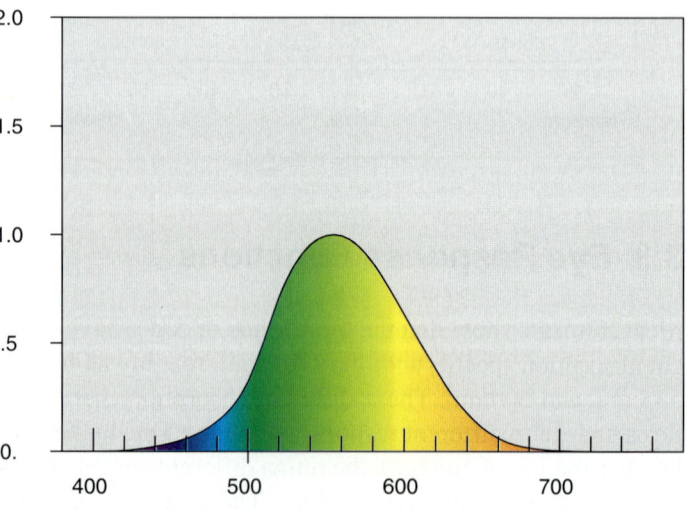

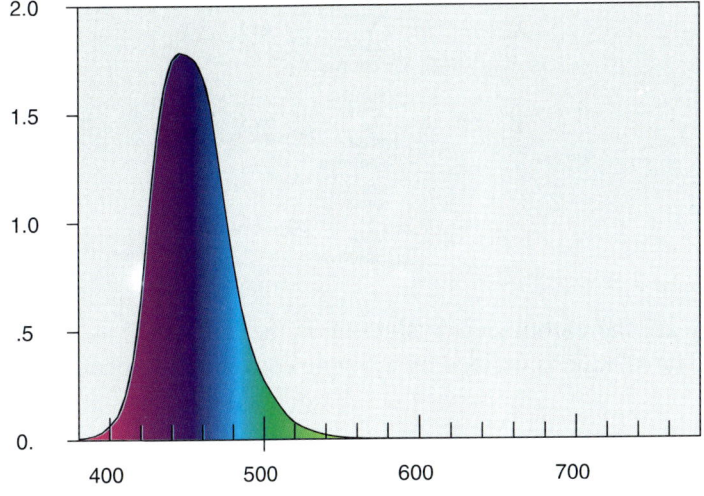

CIE defined in 1931 a standard for measuring perceived color by using these eye response functions $\bar{x}$, $\bar{y}$, and $\bar{z}$ for a 2° standard observer. This standard is called CIE XYZ; the values for X, Y, and Z are computed by integrating the product of the spectral distribution reaching the retina and the response functions as shown in the following formulas. λ is the wavelength, j_λ is the value of the wavelength distribution at λ, and k is a normalization factor such that for pure white (i.e. for the equal energy spectrum) holds $Y_{white} = 100$.

$$X = k \int \varphi_\lambda \bar{x}(\lambda) d\lambda$$

$$Y = k \int \varphi_\lambda \bar{y}(\lambda) d\lambda$$

$$Z = k \int \varphi_\lambda \bar{z}(\lambda) d\lambda$$

X, Y, and Z are called the tristimulus values uniquely defining a color perception.

For practical purposes the integrals are substituted for sums over the range in 5 nm (or sometimes in 10 nm) intervals. This gives the following formulas on the next page.

These absolute values can be normalized using the CIE Yxy color system. This system is also the one most often used for graphical representation of chromaticity values. Here x and

$$X = k \sum_{\lambda_i=380nm}^{780nm} \varphi_{\lambda_i}\overline{x}(\lambda_i)\Delta\lambda_i$$

$$Y = k \sum_{\lambda_i=380nm}^{780nm} \varphi_{\lambda_i}\overline{y}(\lambda_i)\Delta\lambda_i$$

$$Z = k \sum_{\lambda_i=380nm}^{780nm} \varphi_{\lambda_i}\overline{z}(\lambda_i)\Delta\lambda_i$$

y are the chromaticity values, Y is the luminance as above. The relation is defined by a simple conversion mechanism:

$$x = \frac{X}{X+Y+Z}$$

$$y = \frac{Y}{X+Y+Z}$$

Due to its shape the associated diagram 3.10 is also called "shoesole" diagram; the figure gives the xy diagram in the form we will use more often later.

Ill. 3.10:
CIE xy shoesole
diagram

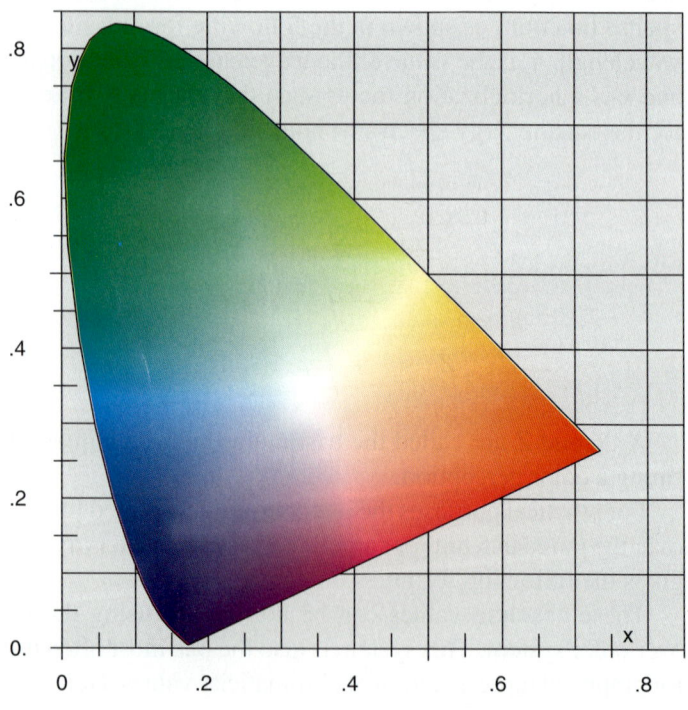

3.4 Illuminants (Daylight, A, B, C, D$_{50}$, D$_{65}$) and White Point

Illumination must be standardized for comparable results in color measurement. Basic for our normal vision is daylight. The typical spectral energy distribution of daylight is shown in diagram 3.11.

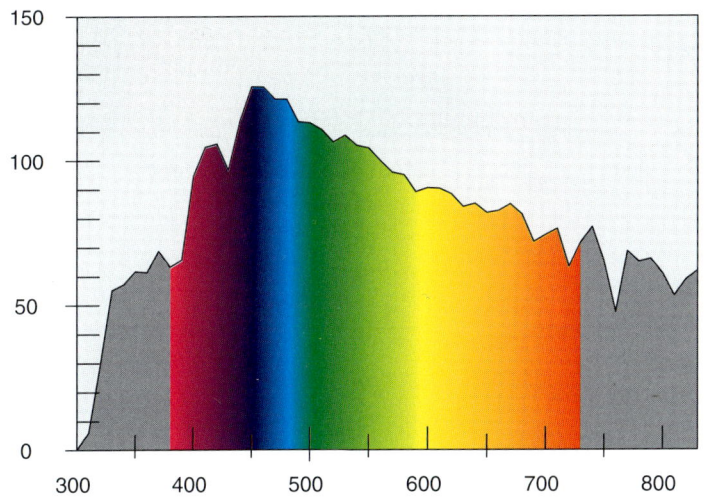

Ill. 3.11:
Typical
spectral energy
distribution of
daylight

Since daylight is not always available in reproducible form, other spectral distributions have been defined as standard illuminants. Illuminants A, B, C, D$_{50}$, and D$_{65}$ with different spectral energy distributions have been standardized; their spectral distribution is shown in the diagrams 3.12–3.16 on pages 16–18.

These different illuminants show significantly different characteristics. Illuminant A for example has a strong dominance of long wavelengths, i. e. it provides a relatively "warm" illumination similar to illuminations seen with normal light bulbs in our houses.

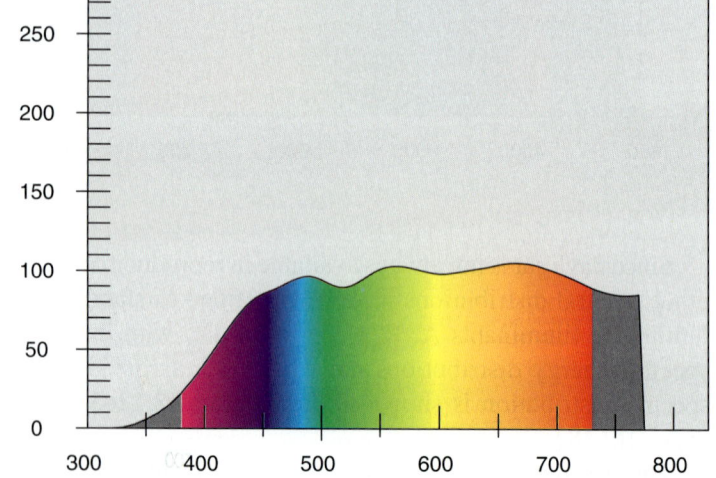

Ill. 3.12:
Spectral energy
distribution of
standard
illuminant A

Ill. 3.13:
Spectral energy
distribution of
standard
illuminant B

D_{50} and D_{65} on the other side are illuminants simulating daylight. As daylight is not constant – noon time daylight shows different characteristics than daylight in the afternoon or close to sunrise or sunset – D type illuminations have been specified according to their color temperature.

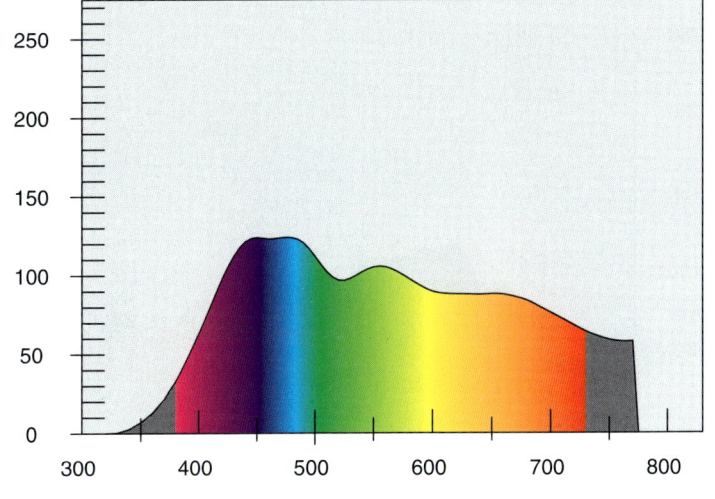

Ill. 3.14:
Spectral energy
distribution of
standard
illuminant C

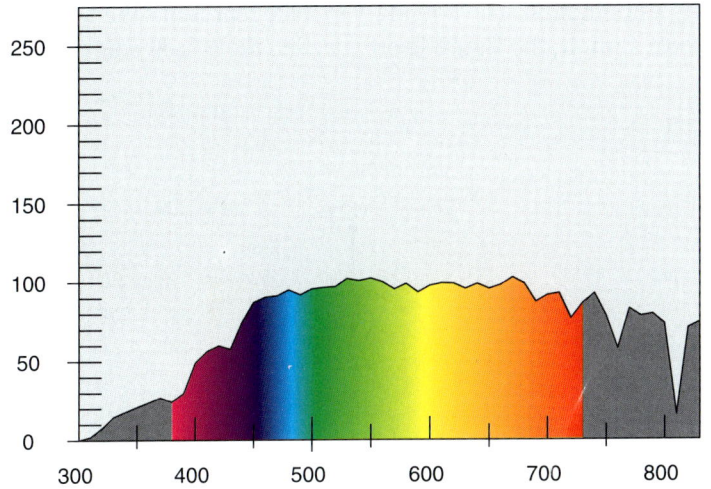

Ill. 3.15:
Spectral energy
distribution of
standard
illuminant D_{50}

So D_{65} stands for an illuminant with approximately 6500° K color temperature, D_{50} stands for approximately 5000° K color temperature. In general, the index of D type illuminants stands for approximately $1/100$ of the color temperature in ° K.

17

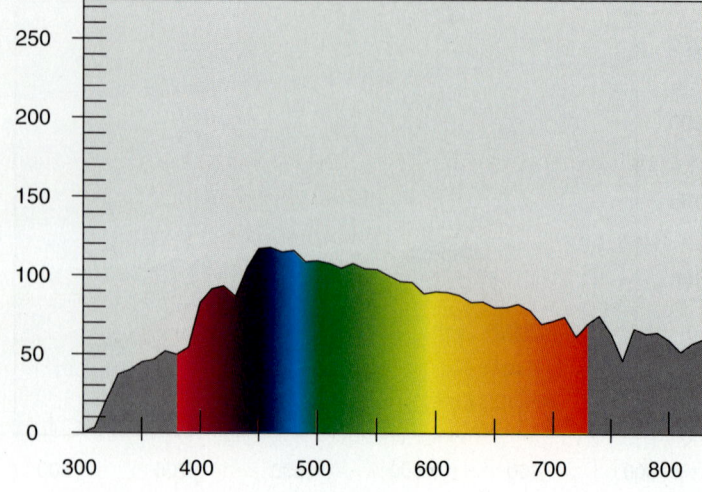

Ill. 3.16:
Spectral energy
distribution of
standard
illuminant D$_{65}$

The diagram 3.17 shows the color space producable under D$_{65}$ illumination. The boundary surface represents the optimal color stimuli; the figure itself is called the Rösch color solid.

Ill. 3.17:
Rösch color solid

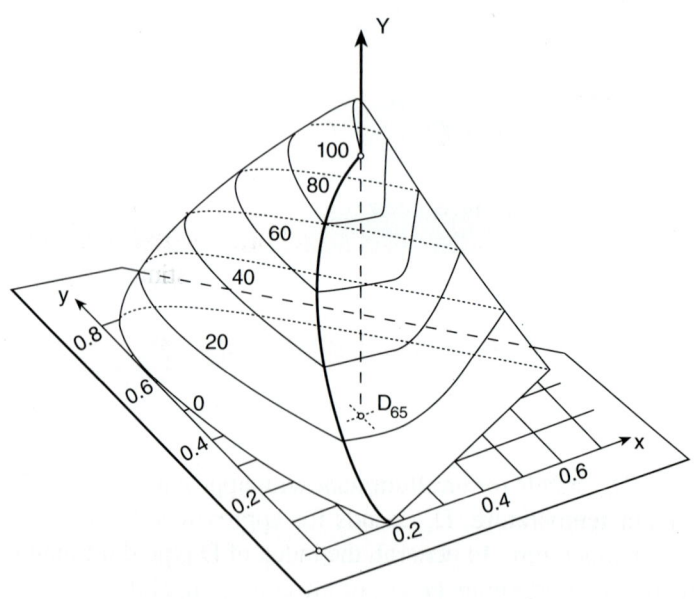

3.5 Metamere Colors

Since the XYZ-values defined by CIE are the values of integrals, one specific value can be reached by numerous different spectral distributions. This fact leads to an effect called metamere colors. This means that colors perceived differently under one illuminant can become identical color perceptions under another illuminant. Missing light energy in one area of the spectrum can be compensated by more light energy in another area of the spectrum; it is only important for metamere colors to preserve the values of the integrals.

This effect is easily demonstrated in everyday life. If you buy pants and a jacket in a shop with artificial illumination, they may match perfectly (be of the same perceived color) under these lighting conditions. If you take these pants and jacket to the daylight, you may experience a surprise: they do not fit well at all. Under the artificial illumination the absorption of the material together with the illumination produced the same color perception, the same XYZ-values; under other illumination conditions the perceived colors were different because of the different spectral distribution of daylight from the artificial illumination.

3.6 Tristimulus vs. Opponent Color Theory

Different theories have been developed to explain the phenomena of color vision. Two major theories are the tristimulus theory and the opponent color theory. The tristimulus theory is based on the three kinds of cones as the primary receptors for color vision. This is in accord with the already mentioned XYZ model of CIE. A color perception can be uniquely identified by the three values X, Y, and Z, and this forms the basis for the theory.

The opponent color theory goes beyond the basic receptors in the retina; it includes combinatorial synapses that take the response of the cones and form signals according to opponent pairs blue-yellow, red-green, and dark-bright. The sketch below illustrates this relationship. There are different theories about how exactly the synaptic relationship is, and this simple figure 3.18 is meant merely to give an impression of the possible synaptic relationships involved. This model also uses three components, however it is not simply the response of the cones, but a color perception derived from additive or subtractive combinations of responses.

Ill. 3.18:
Schematic
diagram of
synapses for
color vision

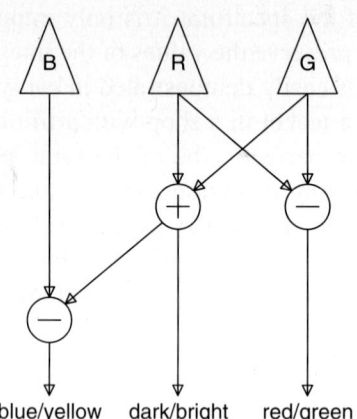

blue/yellow   dark/bright   red/green

Both theories have their merits, both can explain some effects, but it is not fair to say that one theory is superior over the other; both play an important role in explaining color perception. The opponent color theory is based on the tristimulus theory, and they both describe effects at different stages of the perceptive processing of optical stimuli.

3.7 Model of Vision for Colored Objects

We have been looking to color perception without really talking about colored objects. In order to explain the vision of colored objects, we start at the light source where we have a certain energy distribution of light. This light falls on a colored surface; the most important property of the colored surface for our discussion is its ability to reflect some wavelengths of

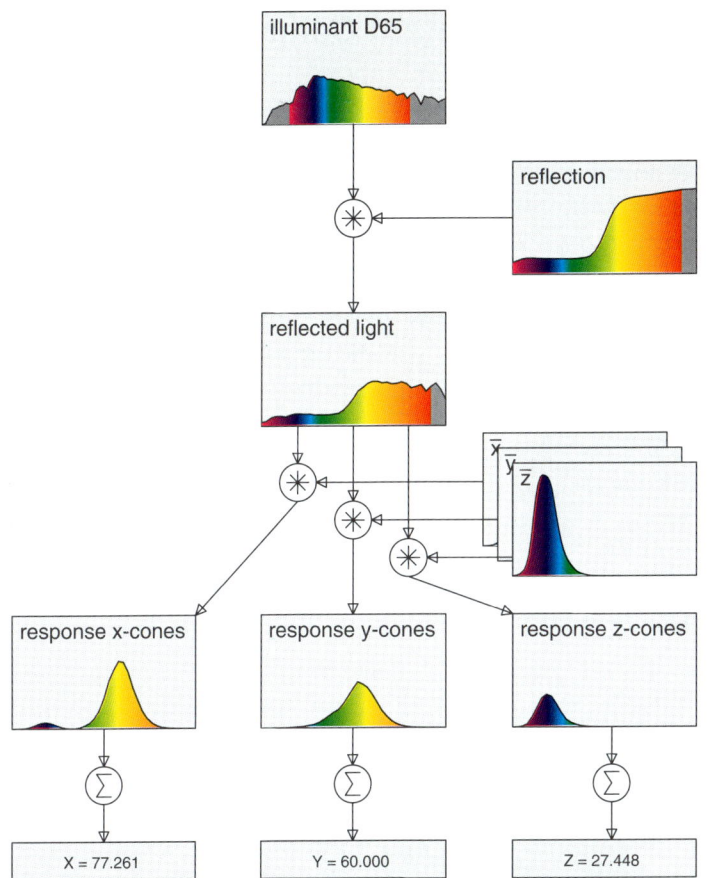

Ill. 3.19: Spectral distribution at various stages of color vision

light while light of other wavelengths is absorbed. So the light emitted by our light source is modified selectively (reflected or absorbed) by the illuminated surface. If we take the reflection (or remission) function $r(\lambda)$ of wavelength λ and multiply it with the energy distribution of the illuminant $e(\lambda)$ we have a measure of the energy distribution of the light reflected from the surface. This energy distribution is used to calculate the perceived color values X, Y, and Z. The diagram 3.19 shows the energy distribution functions of the illuminant, the product of illuminant and remitting function and its product with the respective response functions.

21

3.8 Equally Distant Color Systems

The CIE XYZ color system has nice properties with regard to linearity; if two light sources with certain perceived colors are combined, the resulting perceived color value lies on the linear connection between the two original color values. This property gives a relatively easy way of calibration for color displays. The resulting color value is calculated from the primary color (the color coordinates) of the three phosphors of the tube. However CIE XYZ has bad properties regarding perceived visual distance. The distances of colors that are just distinguishable by the human visual system are very different in different areas of the color space. The differences are illustrated in the diagram 3.20 showing just distinguishable color in form of so-called Mac Adams ellipses magnified by a factor of 10.

In order to provide some perceptual metrics in color distance both linear and non-linear transformations have been

Ill. 3.20:
Mac Adams
ellipses in
CIE xy diagram

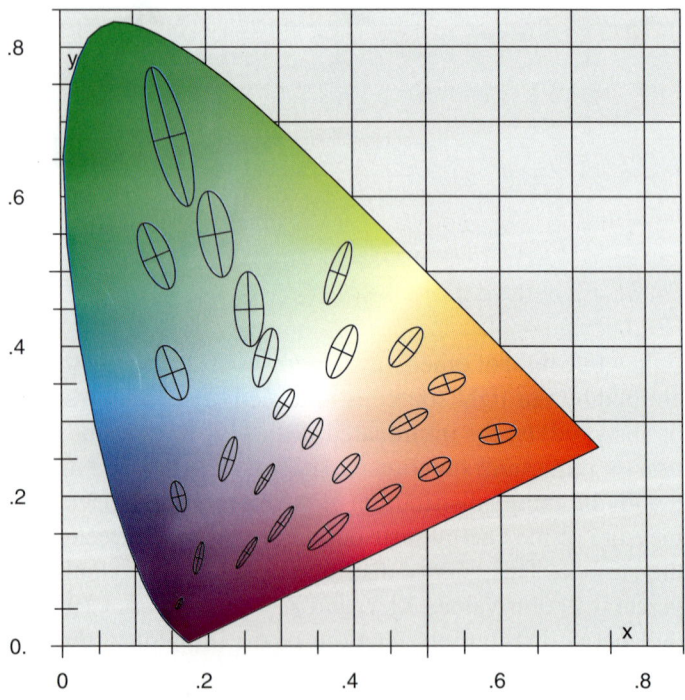

developed deriving (almost) perceptually equal distance color systems which also have been standardized by CIE known under the name CIE L*u*v* and CIE L*a*b*. Both systems are derived from CIE XYZ, however both systems are depending on the definition of white. Both CIE L*a*b* and L*u*v* are examples of perceptionally uniform color spaces. L*u*v* retains its properties regarding linear combination of colors and can therefore also be used to describe the color gamut of a device, while L*a*b* is used especially for color difference measurement, e.g. for describing deviations of colors in printed output.

3.8.1 CIE L*u*v*

The value L* stands for the psychometric lightness, derived from the CIE 1931 luminance value Y, u* and v* are the respective chrominance values. The following formulas define the system:

$$L^* = 116 \left(\frac{Y}{Y_w}\right)^{\frac{1}{3}} - 16, \quad \text{if } \frac{Y}{Y_w} > .008856$$
$$L^* = 903.3 \left(\frac{Y}{Y_w}\right), \quad \text{if } \frac{Y}{Y_w} \leq .008856$$

The symbols X, Y, and Z stand for the respective CIE values, the index w stands for the respective value of the nominal white color stimulus.

Here are the definitions of the u* and v* terms of L*u*v* expressed in CIE XYZ:

$$u^* = 13L^*(u' - u'_w)$$
$$v^* = 13L^*(v' - v'_{...})$$

where:

$$u' = \frac{4X}{X + 15Y + 3Z}$$

$$v' = \frac{9Y}{X + 15Y + 3Z}$$

23

$$u'_w = \frac{4X_w}{X_m + 15Y_m + 3Z_m}$$

$$v'_w = \frac{9Y_w}{X_w + 15Y_w + 3Z_w}$$

3.8.2 CIE L*a*b*

The CIE L*a*b system is an example of an opponent color system. The value L* stands for the psychometric lightness (which is identical to the L* value of the L*u*v* system), derived from the CIE 1931 value Y, a* and b* are the respective chrominance values. The following formulas define the system:

$$L^* = 116 \left(\frac{Y}{Y_w}\right)^{\frac{1}{3}} - 16, \quad \text{if } \frac{Y}{Y_w} > .008856$$
$$L^* = 903.3 \left(\frac{Y}{Y_w}\right) \quad , \quad \text{if } \frac{Y}{Y_w} \leq .008856$$

Again, the symbols X, Y, and Z stand for the respective CIE values, the index w stands for the respective value of the nominal white color stimulus.

Here are the definitions of the a* and b* terms of L*a*b*:

$$a^* = 500 \left[f\left(\frac{X}{X_w}\right) - f\left(\frac{Y}{Y_w}\right) \right]$$

$$b^* = 200 \left[f\left(\frac{Y}{Y_w}\right) - f\left(\frac{Z}{Z_w}\right) \right]$$

where:

$$f\left(\frac{X}{X_w}\right) = \left(\frac{X}{X_w}\right)^{\frac{1}{3}} \quad , \quad \text{if } \frac{X}{X_w} > .008856$$
$$f\left(\frac{X}{X_w}\right) = 7.787 \left(\frac{X}{X_w}\right) + \frac{16}{116} , \quad \text{if } \frac{X}{X_w} \leq .008856$$
$$f\left(\frac{Y}{Y_w}\right) = \left(\frac{Y}{Y_w}\right)^{\frac{1}{3}} \quad , \quad \text{if } \frac{Y}{Y_w} > .008856$$
$$f\left(\frac{Y}{Y_w}\right) = 7.787 \left(\frac{Y}{Y_w}\right) + \frac{16}{116} , \quad \text{if } \frac{Y}{Y_w} \leq .008856$$
$$f\left(\frac{Z}{Z_w}\right) = \left(\frac{Z}{Z_w}\right)^{\frac{1}{3}} \quad , \quad \text{if } \frac{Z}{Z_w} > .008856$$
$$f\left(\frac{Z}{Z_w}\right) = 7.787 \left(\frac{Z}{Z_w}\right) + \frac{16}{116} , \quad \text{if } \frac{Z}{Z_w} \leq .008856$$

The diagram 3.21 shows the relationship of colors in the CIE L*a*b* color coordinate system to the CIE xy chromaticity coordinates.

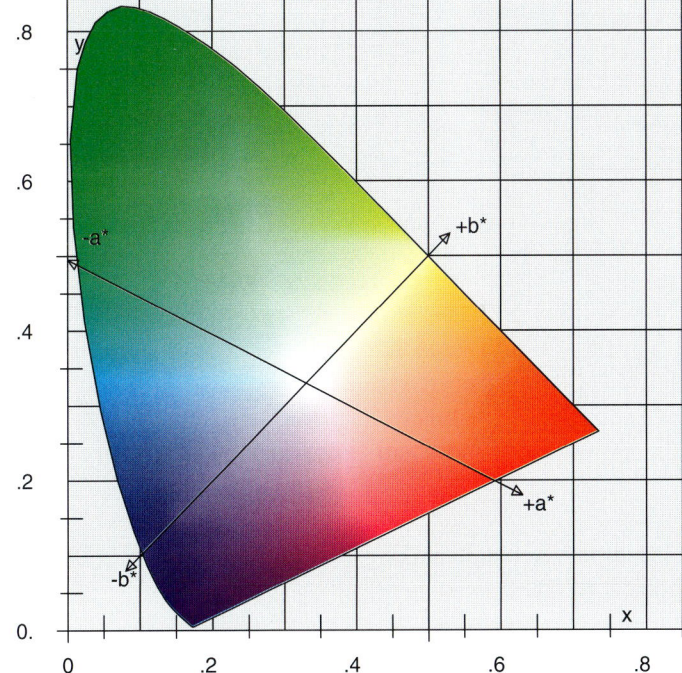

Ill. 3.21:
*Chromaticity coordinate axes of CIE L*a*b* color system mapped to CIE xy diagram*

In diagram 3.22 on page 26 the chromaticity components a* and b* are shown in a rectangular coordinate system at one level of L*; the axis for L* is orthogonal to the page. The opponent colors can be clearly seen from the diagram: negative values of a* are greenish colors, positive values of a* are reddish colors, negative values of b* are bluish colors, positive values of b* are yellowish colors. This diagram also shows in an example the possibility to use saturation (colorfulness) C* and hue h* (as an angle) to specify chromaticity.

3.8.3 Color Difference Measures

Various methods to measure color difference have been proposed; the most prominent one is based on CIE L*a*b* and is

Ill. 3.22:
Chromaticity
coordinates of
the CIE L*a*b*
color system

also approved by CIE. The following formulas show the relations:

$$\Delta E_{ab}^* = \left[(\Delta L^*)^2 + (\Delta a^*)^2 + (\Delta b^*)^2\right]^{\frac{1}{2}}$$

with:

$$\begin{array}{rcl} \Delta L^* & = & L_1^* - L_2^* \\ \Delta a^* & = & a_1^* - a_2^* \\ \Delta b^* & = & b_1^* - b_2^* \end{array}$$

As can be easily seen from the formula this is simply the Euclidean distance in the L*a*b* color space.

Another form measuring color differences uses the differences in lightness L*, saturation C*, and hue h* as in the following equations:

$$\Delta C_{ab}^* = C_{ab_1}^* - C_{ab_2}^*$$

with:

$$C_{ab}^* = \left[(a^*)^2 + (b^*)^2\right]^{\frac{1}{2}}$$

and

$$\Delta h_{ab} = h_{ab_1} - h_{ab_2}$$

with:

$$h_{ab} = arctan(\frac{b^*}{a^*})$$

Indices 1 and 2 in the previous formulas refer to samples 1 and 2 respectively.

The value most often found in classifying print reproduction quality is ΔE_{ab} or short ΔE; its size is a good measure for the quality of a reproduction. ΔE values between 1 and 3 signify an extremely good reproduction and are only achievable with press printing today; typical high quality desk top printers can achieve a ΔE of around 10.

3.8.4 Standard White Values

In order to compute the color coordinates according to the above given formulas it is necessary to specify the respective standard white colors. The white values to be used under standard conditions for different illuminants are given in the following table for a 2° observer.

2° observer	A	B	C	D$_{50}$	D$_{65}$
X_w	109.85	99.09	98.07	96.42	95.05
Y_w	100.00	100.00	100.00	100.00	100.00
Z_w	35.58	85.31	118.22	82.53	108.90

White values for standard illuminants and a 2° observer

The following table gives the respective values for a 10° observer.

10° observer	A	B	C	D$_{50}$	D$_{65}$
X_w	111.14	99.42	97.82	96.72	94.81
Y_w	100.00	100.00	100.00	100.00	100.00
Z_w	35.20	86.45	116.14	81.44	107.34

White values for standard illuminants and a 10° observer

Together with these values the previous formulas can be applied to practical test cases.

3.8.5 Metameric Index

Another important concept for color fidelity of colored surfaces under different lighting conditions is the metameric index. It is a measure for the stability of the color perception under different illuminations.

The metameric index is the perceptional color difference of two objects that are perceived as the same color under one illuminant, under different illumination conditions. The diagrams 3.23–3.25 show the remission curves of three objects that are perceived as equal color under D_{65} illumination. Changing lighting to illuminant A shows significantly different color perceptions.

Ill. 3.23:
Spectral remission curves of three objects with identical color perception under D_{65} illumination for a 2° observer, no. 1

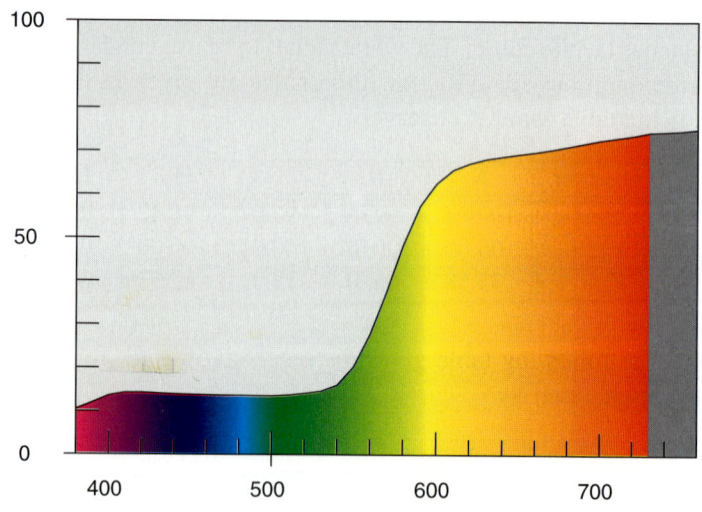

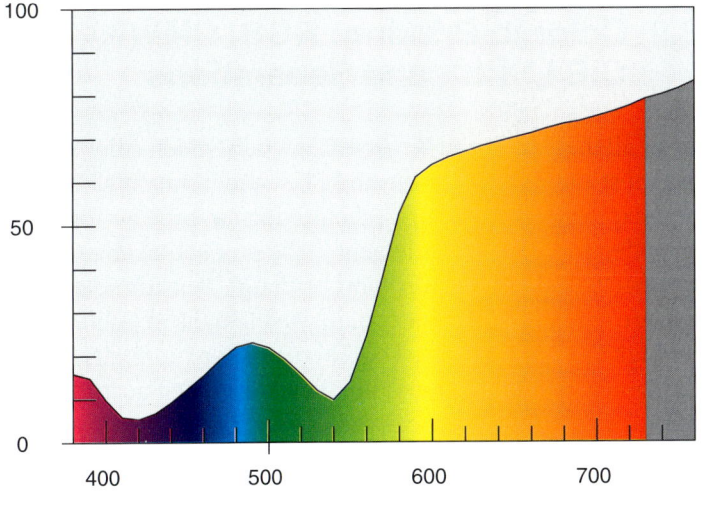

Ill. 3.24:
Spectral remission
curves of three
objects with
identical color
perception under
D_{65} illumination
for a 2° observer,
no. 2

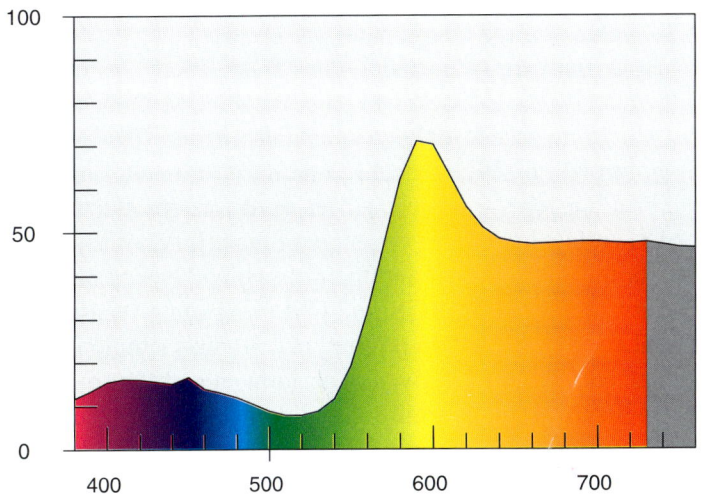

Ill. 3.25:
Spectral remission
curves of three
objects with
identical color
perception under
D_{65} illumination
for a 2° observer,
no. 3

3.9 Color in Printing: RGB, CMY(K), CIE L*a*b*, and Separations

After all these different theoretical models which color spaces are most useful in prepress and press environments? This is best explained by looking at the production process for color reproductions. The first element in the production chain is the scanner. Scanners typically work with light sensitive elements that are made sensitive for certain wavelengths of light mostly through the use of filters. The filters used are normally called red, green, and blue filters, i.e. they are chosen according to the color sensitivity of the human cones. So the primary result of the scanning process are normally RGB data. In most cases this is a device RGB, i.e. it is only in rare exceptional cases a calibrated RGB. If the filters and the lamp to illuminate the input target are properly chosen and adapted to the response behavior of the light sensors, one could easily produce calibrated RGB, i.e. RGB where the primary valences (the CIE XYZ values) of the components are known.

However very often today the obtained RGB values are immediately converted even to process color CMYK data without providing access to the original information. This is common, but bad practice, because one must know at scan time the printing process which will be used for reproduction, as CMYK refers to the area coverage of the printing inks on paper (or the area on film) and their respective dyes to be used. The relation between RGB and CMYK is often regarded as trivial and linear; even in prominent textbooks simple formulas for conversion are given. The form most often found is given here with the warning that it does not reflect the complex situation of light reflected from dyes on a carrier substrate, and that it does not correctly model the process:

$$R = 1 - C$$
$$G = 1 - M$$
$$B = 1 - Y$$

or the inverse:

$$C = 1 - R$$
$$M = 1 - G$$
$$Y = 1 - B$$

with the Black component computed according to the following formula:

$$K = min\,(C, M, Y)$$

In this model often the C, M, and Y values are reduced by the value of K leaving at least one color component with zero value.

Color values range from 0 to 1, where 1 means full intensity for RGB, and 100% area coverage in the case of the primary printing colors C, M, Y, and K. These separation formulas make simplistic assumptions including a linear relationship between RGB and CMY(K). To demonstrate this let us make a simple experiment: printing a sample with a tonal ramp with equal amounts of Cyan, Magenta, and Yellow under these assumptions should produce a wedge with identical chromaticity across the whole area; the difference should only be in the lightness. The chromaticity diagram 3.26 shows the measured curve resulting from overprinting equal amounts of C, M, and Y on an ink jet printer.

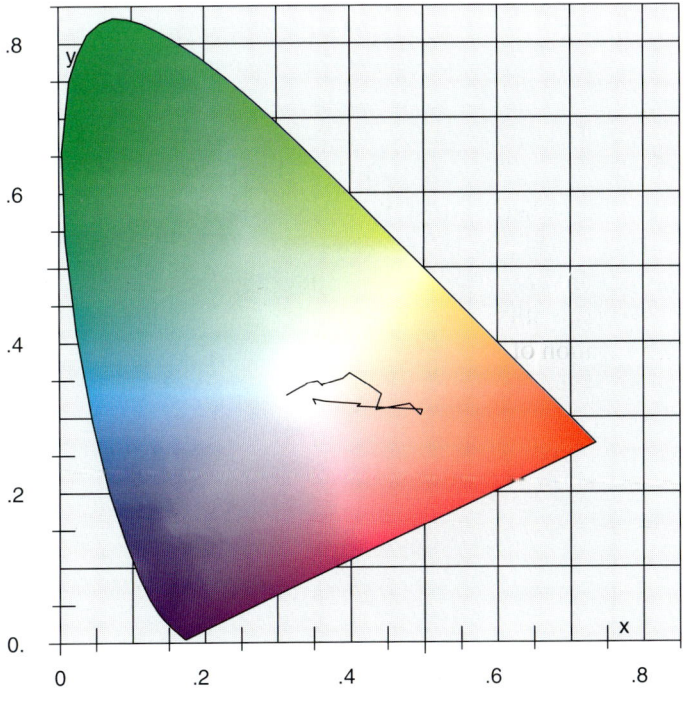

Ill. 3.26:

Chromaticity of overprinting equal amounts of Cyan, Magenta, and Yellow

As this experiment shows, in real life the relationship is highly non-linear; besides having non-ideal dyes, the actual printing results, for example, on an offset press, are influenced by more than 30 parameters. As an example there is a significant difference in the printing results if the process colors are printed in different order; typically the first printed ink shows higher density values, i.e. more ink remains on the paper than inks overprinting other inks.

Besides this problem not always is four color printing sufficient; printing methods are in use that need more separation colors (e.g. seven) to

- produce better color,
- print with a larger color gamut,
- when some special effect is needed, e.g. for overprinting gold, or silver,
- for having some specific spot color to accurately print a certain color that cannot be reproduced accurately enough by the standard process colors, e.g. when a company logo is to be reproduced faithfully whose color cannot be properly matched by using standard process color inks only.

*Ill. 3.27:
Chromaticity
diagram of gamut
of four color
offset print with
standard
European inks*

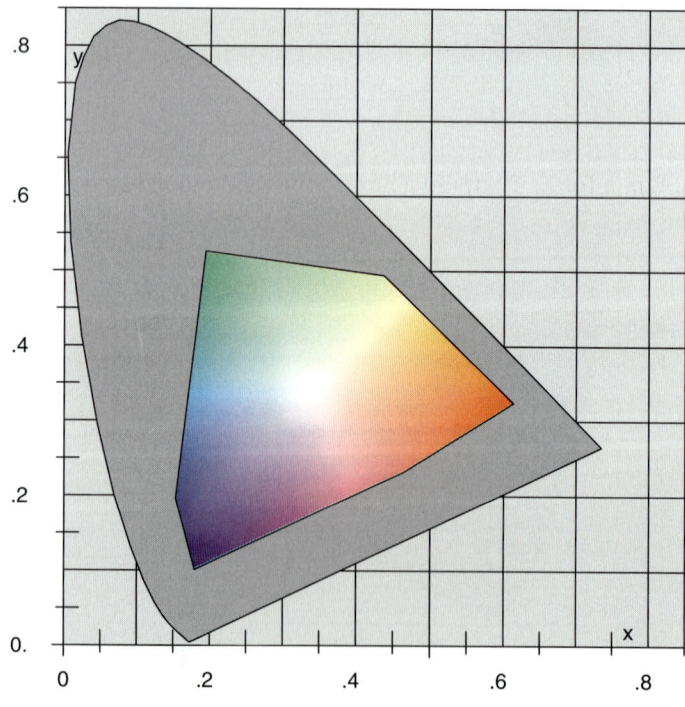

The chromaticity diagram 3.27 on page 32 shows the color gamut achievable on a four color offset press with inks according to the European standard printer inks.

Over the last few years the International Color Consortium (ICC) have undertaken efforts to provide uniform color across devices. The ICC has established guidelines for calibration of devices. The intention is to provide uniform color management as a basic feature of operating systems. These efforts are harmonized with the so-called Color Rendering Dictionaries (CRDs) possible under PostScript Level 2. The representation of the ICC Color Profile is actually done as a tagged binary file where the entries can be easily mapped to a CRD. The issues related to PostScript will be discussed later. The conceptual color rendering pipeline according to PostScript Level 2 gives a good insight in the problems, and is shown in diagram 5.2 in chapter 5.1.5.

Consistent color rendering across devices is attempted by the establishment of a CIE XYZ based color system as central uniform system, which then in turn is rendered with appropriate device rendering procedures provided in the dictionary. Faithful application of this technique at least leads to more accurate results than just applying device color capabilities.

The device color systems that have been around in PostScript before Level 2 are still included. As part of standardization within the ISO TC130 on Graphics Arts (and the related ANSI committee IT8) calibration targets have been defined for both input (scanning) and output. The chart 3.28 on page 34 shows a sample of the printer test form. Regarding output the efforts of the ICC go beyond those of TC130 as they also provide the notion of rendering intent. Rendering intent signifies what is most important to be correctly reproduced; it defines what the term "correctly" means under the assumption that the device gamut is not sufficient to render output exactly; ICC distinguishes the following rendering intents:

- Perceptual: maintains perceptual differences, i.e. the colors are interpreted relative to the device's white point. An important property of this rendering intent is the capability to do gamut compression, i.e. if colors to be presented are outside the device gamut, a gamut compression is performed in order to retain the overall per-

Ill. 3.28:
Example of
an ISO TC130
printer test sheet

ceptual impression of an image. This is typically best for scanned (photographic) images. It is implemented using highly non-linear interpolation.

- Colorimetric: maintains individual colors at the cost of relations between colors that may fall outside the device gamut. This is typically best for spot colors. It is realized by maintaining minimal ΔE values.
- Saturation: maintains levels of color saturation at the expense of contrast. This is best used for computer generated graphics and produces vivid colors.

- Fast match: compromise to make fast rendering possible. This is based on linear color interpolation.
- If none of the above mentioned standard rendering intents provide the desired result, more complex color management methods can be implemented that perform as desired.

3.10 Halftoning

Displays typically allow to present a large variety of tonal values both in gray as well as in color. In printing tonal variation is more limited. Except with some special device technology, e.g. with dye sublimation, only two states per printing dye at one pixel location are available. With two tonal values the simplest approximation at device resolution is to use thresholding to map a range of tonal values to two values only. While this gives an interesting graphic effect it is not able to reproduce images with subtle tonal variations. The examples 3.29 and 3.30 show such a graphic effect.

To provide good comparison the following set of examples show what should be generated, and subsequently, what is generated with the mentioned method.

Ill. 3.29:
Graphic effect
produced with
thresholding

With every pixel in the image we produce a certain deviation from the intended tonal value. This can be compensated by accumulating the error, comparing the error to the threshold, and compensating for that error by omitting or setting a pixel which, if only based on the threshold, would not have been set or omitted.

35

Ill. 3.30:
Graphic effect
produced with
thresholding

The examples 3.31 and 3.32 on page 37 show the effect of such an error distribution method. The common Floyd/ Steinberg algorithm [FL75] is based on the summation of the error and propagates it along each row. This leads to a characteristic propagation pattern which, unfortunately, is clearly visible. Therefore also this method is not a satisfying choice if it comes to quality output of images needing a large range of tonal values.

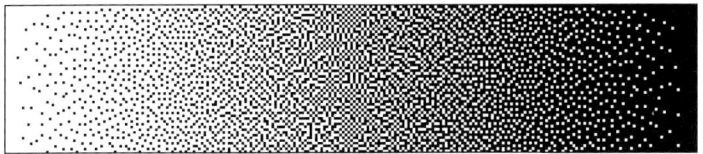

Ill. 3.31:
Error distribution
according to
Floyd/Steinberg

Ill. 3.32:
Error distribution
according to
Floyd/Steinberg

In order to be still able to present a large variety of tonal values, special techniques – called halftoning – have been developed. Halftoning is based on the ability of our eye to integrate light reflected from spots of different dyes over a small area and to give the impression of one color. The presentation of continuous tone images e.g. requires the application of such techniques.

3.11 Simple Model for Halftoning Screens

If the output device only allows bi-level output (ink/no ink) it is necessary to trade tonal resolution for spatial resolution. We do not consider single device pixels but we do combine a region of pixels in order to achieve a specific tonal value. This is mostly done on a cell basis. If we assume a size of 5 by 5 pixels of an axis-parallel square cell, we are able to provide 26 tonal values, 5 times 5 pixels plus no pixel set. In that simple model first order assumption is that for the tonal value only the number of pixels set per area is important. Therefore a trade-off between spatial and tonal resolution is required.

Ill. 3.33:
Sequence of pixels
to be set in cell

13	20	16	23	11
24	7	5	8	19
17	4	1	2	14
21	9	3	6	22
12	25	15	18	10

The simple cell model (see illustrations 3.33 and 3.34) however is not sufficient to describe good halftoning mechanisms, because the distribution of the pixels set in a cell is by no means unimportant for the quality of the reproduction. In our last example we assumed that the cell is square and parallel to the axes. Also this assumption has influence on the tonal output. The illustration 3.35 on page 39 was calculated by applying the screen specified above.

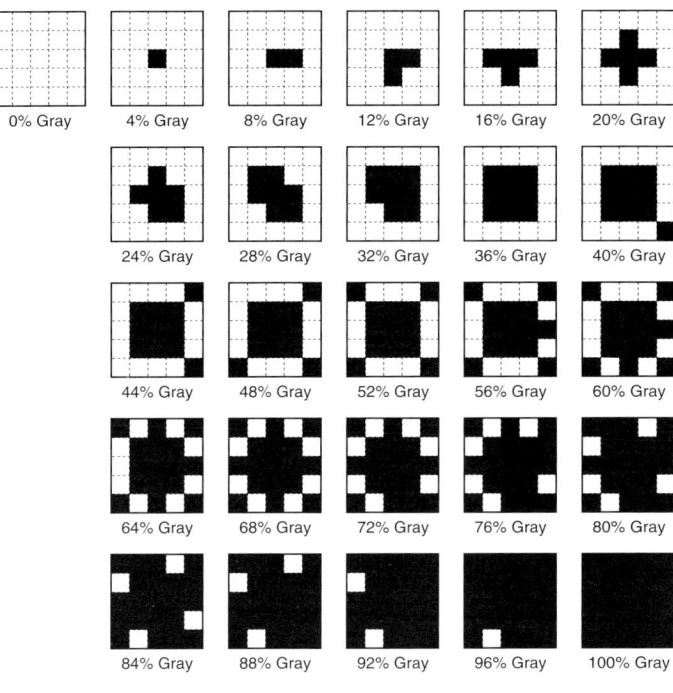

Ill. 3.34:
Example of a cell
model based on a
few tonal values

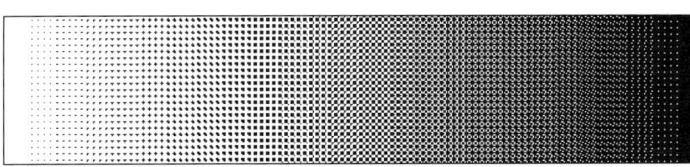

Ill. 3.35:
Example images
based on simple
cell halftoning

39

3.12 Clustered Dot Halftoning

The most common form of halftoning are the so-called clustered dot halftones or screens. They have their name from the fact that the tonal value is produced by dots whose size is manipulated, i.e. pixels are clustered to dots of different size in order to obtain different tonal values. They are also ordered on the printing space under a specific angle, not necessarily parallel to the axes.

A grid of cells is overlaid to the page under a certain (screen) angle and in the cells dots are generated that produce tonal valued output. Since the eye detects regular patterns (like grids) very easily, the cell size has to be small compared to the detail of image to be presented. Due to the regular grid overlaid the distance between dots is always constant, so the frequency (dots per length unit) is also constant. Therefore these screens are also called amplitude-modulated (AM) screens.

Under some angles patterns are more easily visible and disturbing than under other angles. Experiments show that a pattern under an angle of 45° are least visible, the worst situation is when the patterns are parallel to the axes. So it is not surprising that screens for monochrome output are almost always 45° screens. The examples 3.36 and 3.37 (page 41) show a typical 0° and 45° screen.

*Ill. 3.36:
Gray ramp
screened with a
0° AM screen
(top) and a
45° AM screen
(bottom)*

3.13 Dot Shapes

Intuitively one would assume that the best results can be produced using perfectly rotation symmetric shaped dots. In print-

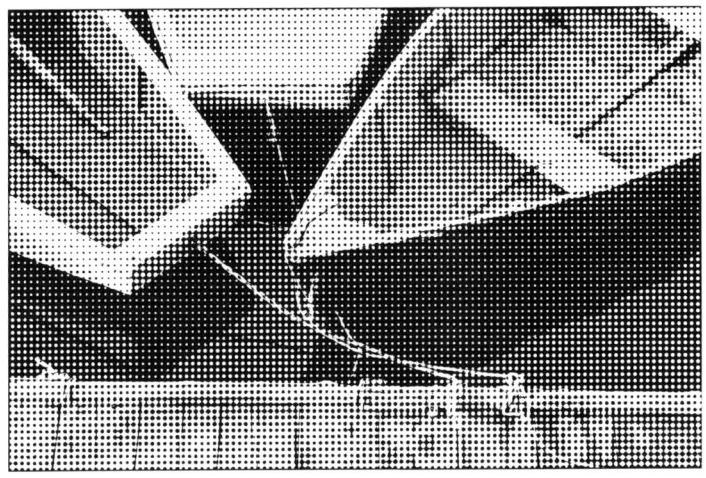

Ill. 3.37:
Example image
screened with a
0° AM screen
(top) and a
45° AM screen
(bottom)

ing however rotation symmetric screen dots produce a visible tonal step in middle tonal values. Rotation symmetric dots shapes have the property of connecting the dots in both grid directions at the same tonal value, an effect called dot join. At the positions where dots connect, there is typically an increase of the inked area due to ink flowing together under pressure. This leads to a significant change of inked area between two adjacent tonal values producing a visible step on the tonal ramp. Around the middle tone values there is a significant discontinuity in dot gain with rotation symmetric dot shapes. There-

41

Ill. 3.38:
Gray ramp
screened with
rotation symmetric
dot shape (top)
and elliptic dot
shape (bottom)

Ill. 3.39:
Example image
screened with
rotation symmetric
dot shape (top)
and elliptic dot
shape (bottom)

fore dot shapes in the middle tone values typically are of elliptic shape, and connecting dots appear at different levels of tone in the two grid axis directions.

The examples 3.38 and 3.39 on page 42 show the effect using screens with rotation symmetric and elliptic dot shapes in the middle tonal values.

The effects shown influence the selection of screening dot functions used for printing presses. Another important property is the number of tonal values that can be presented using a certain screen. Here the number of pixels available in one cell is of major importance. The number of pixels available is depending on the frequency of the screen and the device resolution. If the number of tonal values is not satisfactory, normal screening can be used together with some method of error distribution; this can cause a large increase in the number of tonal values.

3.14 Overprinting of Multiple Separations

In order to achieve output for continuous tone color, multiple color separations have to printed over each other. Overprinting multiple screened separations however confronts us with the typical unwanted moiré effects. These effects are highly correlated with the angle difference of the screened separations. Using three separations, e.g. C, M, and Y, allows the theoretical optimum with the minimal interference patterns with identical frequencies for all separations, the equal distance of the three angles of 30°, placing the darkest component at 45°, and offsetting the other two separations to 15° and 75° respectively. The classical screening methods use equal screen ruling or frequency for all separations. Minor deviations from these angles and frequencies can give annoying moiré effects. Neither the required angles nor the equal frequencies can be produced with simple methods.

With normal process color printing using four color separations the situation becomes even more complicated and critical. The ideal angle difference that works for three separations does not give a solution for four separations. Here the solution most frequently used is applying the method of equal angular distance for the darker separation colors C, M, and K; the Y component (which is the one with the least contrast) is printed parallel to the axes under 0°, and for visual smoothness also with a somewhat larger frequency than the other separations.

Using repeated cell patterns does not allow to exactly produce the angles, and gives different frequencies for each separation. The example 3.40 shows the visual effect of overprinting by overlaying two rectangular grids at different angles.

Ill. 3.40:
Moiré effect from
overprint of grids
at different angles

3.15 Rational vs. Irrational Screens

In order to reduce moiré effects in print new methods had to be found. Here comes the distinction between so-called rational and irrational screens into play.

The screens discussed before have all been of the type rational screen. Rational screen is somewhat a misnomer, because the screen is not rational, but the tangent of the angle is a ratio of integers in terms of the device grid units of the output device.

The goal here really is to approximate the correct angles with an error that is small enough that unwanted visual effects (moiré) disappear or are at least minimized. This can be done by different methods.

One method is to sample an accurate function over space to generate the proper distribution of the screen dots and thereby approximate the angle more correctly.

Another way of achieving more accurate approximation of angles and still use repeatable patterns to generate the screen is the technique of super-cells. Instead of defining a single cell and repeating it all over the page, a super-cell containing many screen cells is applied, and thereby approximating the angle more exactly. The super-cells are computed according to the first method.

A detailed discussion of the PostScript accurate screens method can be found in [FI92].

3.16 FM screens

All screening methods have one property in common: they all produce the tonal value by painting a portion of an area with ink while another portion is left free (i.e. white on white paper). AM screens cluster the inked portion of the area around the screen dots, and screen dots are equally spaced on a regular grid over the area. The tonal value is directly controlled by the size of the screen dots in relation to the elementary grid size. Frequency modulated (FM) screens, to the contrary, do not cluster the ink in dots on a regular grid but rather use dots of one size and place them statistically; here the density of the placed dots decides about the tonal value.

FM screens typically show a more detailed reproduction of small objects, but also exhibit a grainy structure in large areas of small tonal change.

3.17 AM, FM, and Dot Gain

An important aspect for using screens is the ability to reproduce the whole range of tonal values with good accuracy. The number of pixels set in an area in relation to the number of pixels in that area are regarded as a measure for the tonal value produced. The relation between this ratio and the tonal value however is not linear. During printing, due to different effects, an increase of the area carrying ink can be observed.

This effect is called dot gain, because it results in an increase of the dot size of AM screen dots. Many parameters from the printing process influence the dot gain, its main effect is the spread of ink from the ink carrying region to the regions normally not carrying ink. This means the effect will be more noticeable if there are longer borderlines between ink carrying and no ink carrying regions.

This is a theoretical model which matches quite nicely the effects observed in practice. With this model it can be predicted that FM screens are much more subject to dot gain than AM screens. This is also what is revealed in practical printing tests. The model exhibits small dot gain in low and high tonal values, and increased dot gain in the middle tonal range, and it is much more noticeable in FM screens, than in AM screens, because there are more unconnected pixels, and hence longer borderlines.

3.18 A Simple Printing Model: The Neugebauer Equations

Theoretical models for the color printing process are quite rare. There is only one important model that makes the attempt to explain the effects of color printing to some degree. This model is described by the Neugebauer Equations after Hans E. J. Neugebauer, one of the pioneers in the area of color reproduction. His work dates back in the 1930s. Neugebauer´s PhD thesis was published in 1935 under the title "Zur Theorie des Mehrfarbendruckes" [NE35]. It was originally based on three process colors (Cyan, Magenta, and Yellow) only, but has later been extended to also cover four (and more) process color printing [NE37, NE37a, NE49]. The results of this work are summarized by the following three equations (pages 47 and 48) separating the effects according to the CIE XYZ model in X, Y, and Z.

A stands for the area fraction (between 0 and 1), and the index gives the printing ink for that area. The X, Y, and Z values are

$$
\begin{aligned}
X = \;& (1-A_c) \times (1-A_m) \times (1-A_y) \times (1-A_k) \times X_w \; + \\
& A_c \times (1-A_m) \times (1-A_y) \times (1-A_k) \times X_c \; + \\
& (1-A_c) \times A_m \times (1-A_y) \times (1-A_k) \times X_m \; + \\
& (1-A_c) \times (1-A_m) \times A_y \times (1-A_k) \times X_y \; + \\
& (1-A_c) \times (1-A_m) \times (1-A_y) \times A_k \times X_k \; + \\
& A_c \times A_m \times (1-A_y) \times (1-A_k) \times X_{cm} \; + \\
& A_c \times (1-A_m) \times A_y \times (1-A_k) \times X_{cy} \; + \\
& A_c \times (1-A_m) \times (1-A_y) \times A_k \times X_{ck} \; + \\
& (1-A_c) \times A_m \times A_y \times (1-A_k) \times X_{my} \; + \\
& (1-A_c) \times A_m \times (1-A_y) \times A_k \times X_{mk} \; + \\
& (1-A_c) \times (1-A_m) \times A_y \times A_k \times X_{yk} \; + \\
& A_c \times A_m \times A_y \times (1-A_k) \times X_{cmy} \; + \\
& A_c \times A_m \times (1-A_y) \times A_k \times X_{cmk} \; + \\
& A_c \times (1-A_m) \times A_y \times A_k \times X_{cyk} \; + \\
& (1-A_c) \times A_m \times A_y \times A_k \times X_{myk} \; + \\
& A_c \times A_m \times A_y \times A_k \times X_{cmuk}
\end{aligned}
$$

$$
\begin{aligned}
Y = \;& (1-A_c) \times (1-A_m) \times (1-A_y) \times (1-A_k) \times Y_w \; + \\
& A_c \times (1-A_m) \times (1-A_y) \times (1-A_k) \times Y_c \; + \\
& (1-A_c) \times A_m \times (1-A_y) \times (1-A_k) \times Y_m \; + \\
& (1-A_c) \times (1-A_m) \times A_y \times (1-A_k) \times Y_y \; + \\
& (1-A_c) \times (1-A_m) \times (1-A_y) \times A_k \times Y_k \; + \\
& A_c \times A_m \times (1-A_y) \times (1-A_k) \times Y_{cm} \; + \\
& A_c \times (1-A_m) \times A_y \times (1-A_k) \times Y_{cy} \; + \\
& A_c \times (1-A_m) \times (1-A_y) \times A_k \times Y_{ck} \; + \\
& (1-A_c) \times A_m \times A_y \times (1-A_k) \times Y_{my} \; + \\
& (1-A_c) \times A_m \times (1-A_y) \times A_k \times Y_{mk} \; + \\
& (1-A_c) \times (1-A_m) \times A_y \times A_k \times Y_{yk} \; + \\
& A_c \times A_m \times A_y \times (1-A_k) \times Y_{cmy} \; + \\
& A_c \times A_m \times (1-A_y) \times A_k \times Y_{cmk} \; + \\
& A_c \times (1-A_m) \times A_y \times A_k \times Y_{cyk} \; + \\
& (1-A_c) \times A_m \times A_y \times A_k \times Y_{myk} \; + \\
& A_c \times A_m \times A_y \times A_k \times Y_{cmuk}
\end{aligned}
$$

the measured values of a patch overprinted with solid coverage of the inks indicated in the indices. The index w stands for paper white.

The three equations rely on assumptions of the fractions of the area covered by each of the possible ink combinations C, M, Y, K, CM, CY, CK, MY, MK, YK, CMY, CMK, CYK,

$$
\begin{aligned}
Z = \ &(1-A_c) \times (1-A_m) \times (1-A_y) \times (1-A_k) \times Z_w && + \\
&A_c \times (1-A_m) \times (1-A_y) \times (1-A_k) \times Z_c && + \\
&(1-A_c) \times A_m \times (1-A_y) \times (1-A_k) \times Z_m && + \\
&(1-A_c) \times (1-A_m) \times A_y \times (1-A_k) \times Z_y && + \\
&(1-A_c) \times (1-A_m) \times (1-A_y) \times A_k \times Z_k && + \\
&A_c \times A_m \times (1-A_y) \times (1-A_k) \times Z_{cm} && + \\
&A_c \times (1-A_m) \times A_y \times (1-A_k) \times Z_{cy} && + \\
&A_c \times (1-A_m) \times (1-A_y) \times A_k \times Z_{ck} && + \\
&(1-A_c) \times A_m \times A_y \times (1-A_k) \times Z_{my} && + \\
&(1-A_c) \times A_m \times (1-A_y) \times A_k \times Z_{mk} && + \\
&(1-A_c) \times (1-A_m) \times A_y \times A_k \times Z_{yk} && + \\
&A_c \times A_m \times A_y \times (1-A_k) \times Z_{cmy} && + \\
&A_c \times A_m \times (1-A_y) \times A_k \times Z_{cmk} && + \\
&A_c \times (1-A_m) \times A_y \times A_k \times Z_{cyk} && + \\
&(1-A_c) \times A_m \times A_y \times A_k \times Z_{myk} && + \\
&A_c \times A_m \times A_y \times A_k \times Z_{cmyk} &&
\end{aligned}
$$

CMYK, and paper white. These assumptions are based on statistical considerations, and therefore are idealistic and not accurate in practice. Still the Neugebauer Equations give the best approximation currently available to model the process of overprinting multiple color separations.

3.19 Printing Processes on an Offset Press

After examining a possible theoretical model for offset color printing we now look at some practical effects appearing during the process of producing a print product using an offset press. A brochure on color and quality by Heidelberger Druckmaschinen AG [HE95] is a good reference for that kind of problems, and this chapter makes use of the material provided there.

Up till now we have been considering printing as an atomic action, but now we split this process into separate steps or components. These steps may not be separately visible externally during the production; still they exist, and it is important to understand in which specific component which errors are introduced, and how they combine increasing or decreasing the overall error made during the production. Details of an offset press are described in chapter 4.9 *(Presses)*.

3.19.1 Thickness of Ink Coating

A major point of influence in any printing process is the thickness of the ink coating on the paper. Normally a certain nominal thickness is attempted to be achieved which is measured by density measurement. This is required to maintain a reproducible process during a high volume press run, and also in cases where a product is reprinted another time. Consistent quality is an important business criterion, and there is no consistent quality if there is no consistent density measurement, and thus consistent thickness of ink coating.

Given normal offset conditions with art paper and process colors, good results are achieved with a coating thickness between 0.7 and 1.1 microns. For process technical reasons the thickness can maximally be about 3.5 microns. The importance of the correct coating thickness can be deduced from the fact that process inks are transparent and require a certain thickness to develop filter functionality.

3.19.2 Raster Points in the Process

We have seen in the previous discussions about screening how important a factor this is to achieve good quality print. We have considered the screening process by looking at the pixels set at the device level, and at the printed results, comparing the output, and compensating for deviations by transfer functions and color transformations. The history of the raster however includes many intermediary stages between the device pixel on an image setter and the resulting image on paper.

3.19.2.1 Raster Points on Film

Prior to outputting raster on film there is at least conceptually a page buffer; its conceptual pixels are typically square (and in some rare cases rectangular). Pixels on film are never square, they are mostly round, or perhaps elongated or elliptical. The differences in the area covered by those round pixels compared to the area covered by the idcalized square pixels is compensated for by the transfer function – this however is not the

49

slightly). These effects of course have to be controlled (see 3.19.7 *Color Control Strip*). They often involve problems with the rubber sheet.

3.19.4 Printing Characteristics

A difficult task to be solved during printing is to keep linearity. Since the processes involved are not linear, and in order to achieve linearity, nonlinear effects have to be compensated for. One effect of primary importance is the so-called dot gain. Dot gain is the difference between tonal values on the film and in print. These differences are caused by geometric changes in the raster dot (as was explained earlier), and by the so-called light gathering or light trap.

Like tonal value dot gain is expressed in percent. Since the dot gain is different for different tonal values, it is related to the corresponding tonal value of the film, e.g. 15% dot gain at 40% tonal value.

Dot gain can be easily visualized and directly used in form of the so-called print characteristic, the tonal value with dot

*Ill. 3.41:
Schematic
diagram of
light gathering*

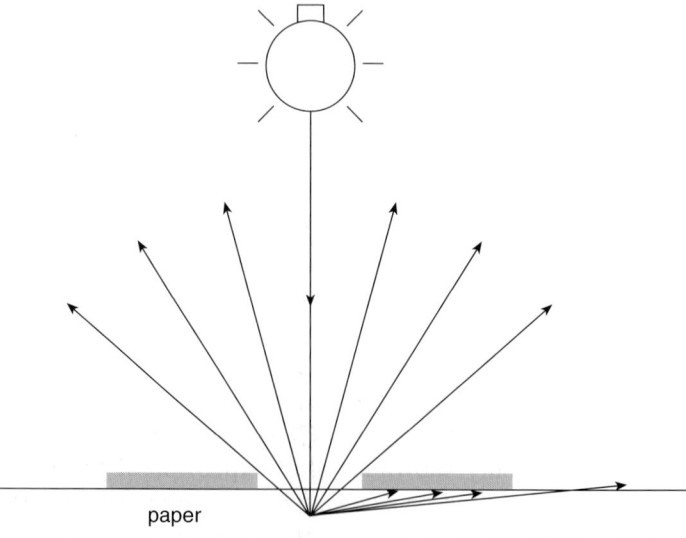

paper

area coverage (%) in print

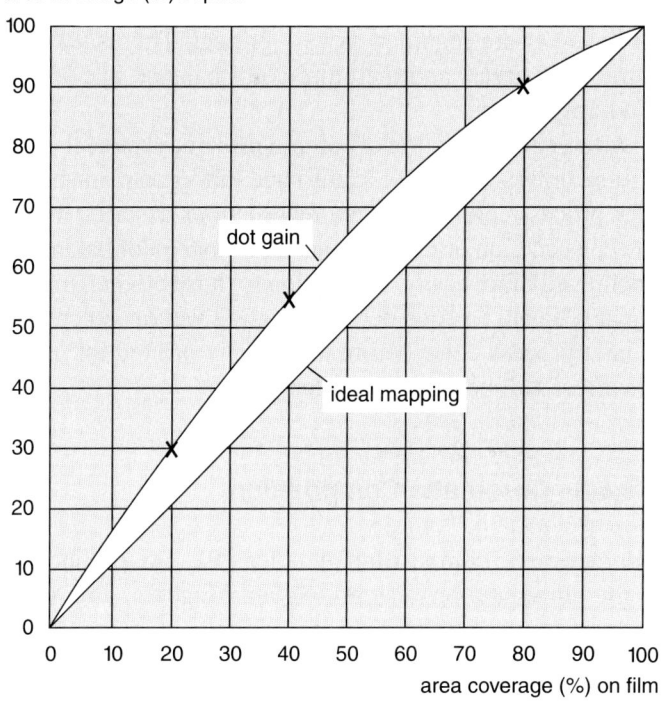

Ill. 3.42:
Print
characteristic

area coverage (%) on film

gain (in print) as a function of the tonal value in film. A characteristic curve is shown in the figure 3.42.

In order to produce correct result in print a pre-compensating function has to be applied before film generation.

3.19.5 Color Balance

Color tones are reproduced in four color print by specific amounts of the four process colors. Changes in these amounts cause changes in the color tones. To obtain reproducible results and stable color reproduction, the supply of the four process colors have to be kept in balance.

Changes in the Black component are regarded as relatively uncritical by the human viewer. Similarly changes of all chromatic color components in the same direction (increase or decrease) is also regarded as not very critical. Hue changes

53

due to changes of chromatic components in different direction however are regarded as very critical. This is visible in neutral gray areas, and therefore color balance is also often called gray balance.

Different construction of color in print may make the process more or less vulnerable against hue shifts due to misbalance of the process colors. The next few sections describe the possible construction principles, and show how color balance can be achieved more easily. In addition to those effects, ink trapping also has to be considered if printing wet in wet (printing the next process color before the already printed ink is dry). This effect will be described later.

3.19.5.1 Chromatic Composition

With chromatic composition all color tones are produced using the chromatic process inks Cyan, Magenta, and Yellow. Black is only used to improve reproduction of image shadows and to support contours. Dark tonal values are produced by an

Ill. 3.43:
Chromatic
composition

70 %	85 %	95 %	0 %	250 %

Ill. 3.44:
Ink relations
with chromatic
composition

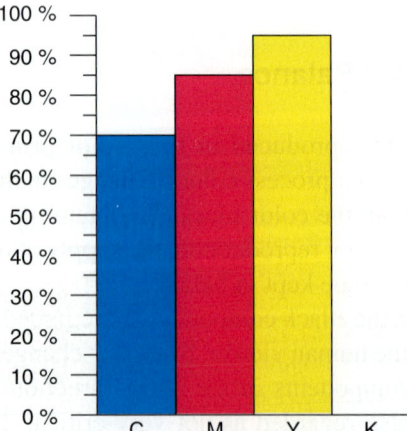

appropriate mix of the three chromatic process colors. The examples 3.34 and 3.44 gives the composition of a dark brown by using only chromatic inks and no Black at all. It is constructed by using an area coverage of 70% Cyan, 85% Magenta, and 95% Yellow. This amounts in much ink in one spot. The diction here is that there is an area coverage of 250% (i.e. the sum of 75, 85, and 95%). Because of the high amount of ink deposited on the printing substrate (e.g. paper) it is very difficult to keep the printing process stable. In addition, as the time to dry a print increases, so does the cost due to the high ink consumption.

3.19.5.2 Under-Color Removal

Under-color removal (UCR) or under-color reduction is a technique that uses chromatic composition, but a part of the three overprinted chromatic inks is substituted by Black ink. Using the same color as in the example for chromatic composition above we could remove 30% of the ink areas of Cyan, Ma-

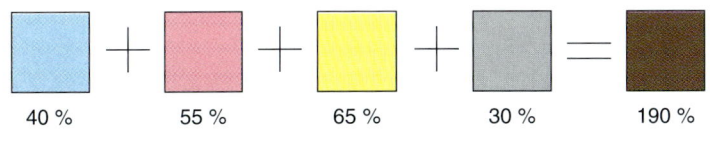

| 40 % | 55 % | 65 % | 30 % | 190 % |

Ill. 3.45:
Chromatic
composition with
under color
removal

Ill. 3.46:
Ink relations
with under color
removal

genta and Yellow and substitute it for Black. This would result in 40% Cyan, 55% Magenta, 65% Yellow, and 30% Black, as shown in the following example. As a consequence the result is an area coverage of 190%, as compared to the 250% of the completely chromatic composition.

3.19.5.3 Achromatic Composition

Achromatic composition removes the achromatic part, i.e. the parts of equal amount of the chromatic colors Cyan, Magenta, and Yellow, completely and substitutes it for Black. This technique is also called gray component replacement (GCR), as the part of neutral gray is completely replaced by Black. This results in 0% Cyan, 15% Magenta, 25% Yellow, and 70% Black, as shown in the following example. As a consequence this would only result in an area coverage of 110%, as compared to 190% in the chromatic composition with under color removal and 250% under completely chromatic composition.

Ill. 3.47:
Achromatic
composition

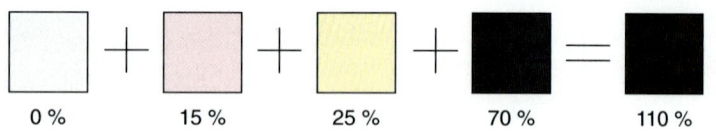

Ill. 3.48:
Ink relations
with achromatic
composition

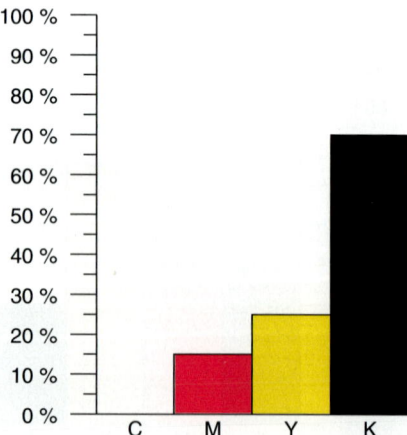

3.19.5.4 Achromatic Composition and Chromatic Color Addition

Chromatic color addition is a commonly used technique. To support image shadows especially in the neutral grays equal amounts of Cyan, Magenta, and Yellow are added on top of an achromatic composition. This method leads to a very stable printing process and produces good quality. Computationally this leads to identical results as under color removal (in this example uncer color removal of 45 %); conceptually chromatic color addition starts from an achromatic composition, while under color removal starts from a chromatic composition.

3.19.5.5 Five/Six/Seven Color Print

In order to extend the printable color gamut, or to produce special effects, more than four inks can be used in print. For example it is possible to use red, green, and blue inks in addi-

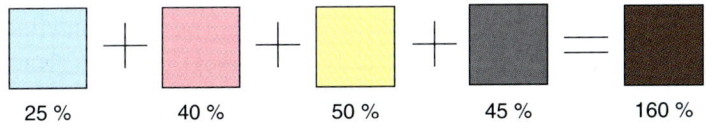

25 %　　40 %　　50 %　　45 %　　160 %

Ill. 3.49: Achromatic composition with chromatic color addition

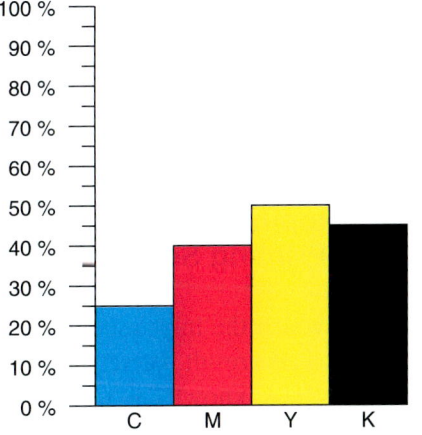

Ill. 3.50: Ink relations with achromatic composition with chromatic color addition

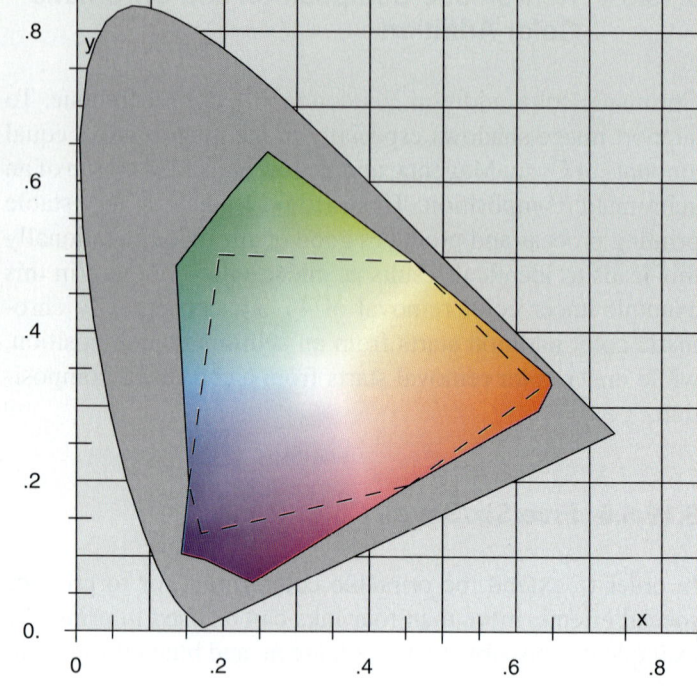

Ill. 3.51:
CIE xy diagram
showing typical
gamuts of four
(hexagon) and
seven (dodecagon)
color print

tion to Cyan, Magenta, Yellow, and Black to extend the gamut significantly towards red, green, and blue. The xy chromaticity diagram 3.51 shows the differences that can result from using seven colors in print compared to standard four color printing.

The inner hexagon shows a typical gamut of four color print; the outer dodecagon shows the respective gamut of a seven color printing process.

Other uses of more than four inks are to reproduce colors that are unreproducible in a standard four color process, for example silver, gold, and other metallic colors.

3.19.6 Ink Trapping and Color Printing Sequence

The sequence in which the inks are applied can make a significant difference in the result. Therefore, the separations have to be prepared with the knowledge of the printing sequence. The reason for this lies in the fact that ink adhesion is better on paper than on top of an already printed ink, if printing wet-on-wet. This effect is called (ink) trapping. It should not to be confused with the term trapping, also called spreading, which means extending an area at the borders to avoid white areas which become visible if the separations are not properly registered. The following diagram illustrates the effects of ink trapping under different print order. The top row shows the realized process; the second row shows the effects of first printing Magenta and then printing Cyan, the third row reverses the print order to Cyan, and then Magenta.

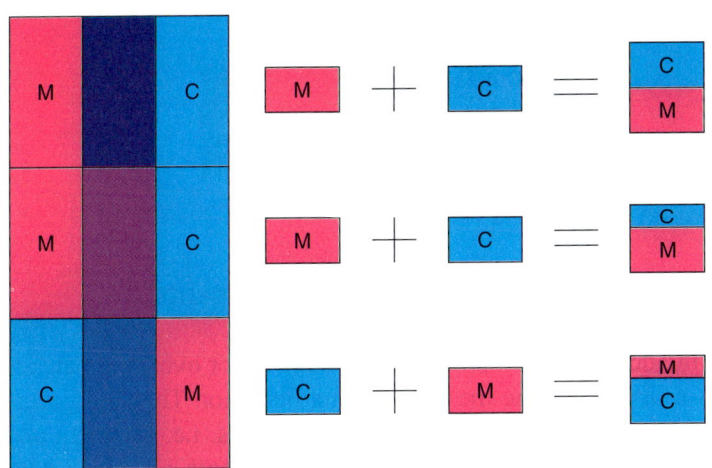

Ill. 3.52: Schematic diagram illustrating ink trapping

3.19.7 Color Control Strip

In order to control quality and maintain a controlled environment, conventional printing uses color control strips. These color control strips are supplied by different vendors as film strips that are copied to the plate together with the page film. In the past to avoid differences, original control strips had to be used. With digital production, when there may not even be a film stage, conventional color control strips of this type are no longer adequate.

Digital color control strips have to be developed. However, most of the control elements are still needed, but special care has to be taken to program these strips in PostScript as the effects produced here normally can only be generated when knowing exact device characteristics.

In the following sections the most important control elements are described. These control elements are also needed in fully digital environments.

3.19.7.1 Full Tone Elements

Full tone elements are simple to produce; these are patches (squares or rectangles) in full tone process colors. In the example B stands for Black instead of the more often used K. These patches are needed to control density of the ink coating.

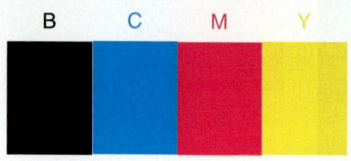

Ill. 3.53:
Full tone
control elements

3.19.7.2 Solid Color Overprint Elements

In addition to full tone patches, solid color overprint patches are used to judge the ink trapping behavior. They are generated by overprinting pairs of ink patches, i.e. Magenta and Yellow, Cyan and Yellow, Cyan and Magenta, and Cyan, Magenta, and Yellow.

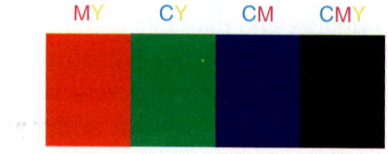

Ill. 3.54:
Solid color
overprint control
elements

3.19.7.3 Color Balance Elements

Color balance elements appear in two versions, as full tone patches, and as raster patches. With full tone patches Cyan, Magenta, and Yellow are overprinted full tone, next to a Black patch. Ideally the first approximately produces a neutral Black.

The raster patches for color balance should give, under controlled print conditions (right ink coating thickness, standard color sequence, and normal dot gain), a neutral gray; the patch is printed next to a Black raster patch with the same tonal value.

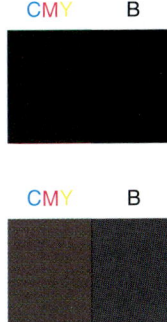

Ill. 3.55:
Full tone and
raster color
balance control
elements

3.19.7.4 Single Color Raster Elements

Single color raster elements form an important element together with the full tone patches to measure dot gain. The elements are implemented quite differently depending on the manufacturer. Heidelberger Druckmaschinen e.g. uses a 70% raster element with round dots; FOGRA uses two elements with 40% and 80% respectively, whereas Brunner uses three different elements, two with 50% and one with 75% area coverage. FOGRA and Brunner are both suppliers of control elements on film. The measuring process has to be adjusted to measuring elements used.

3.19.7.5 Slur and Doubling Control Elements

Line rasters under different angles are used for visual and measured control of slur and doubling effects. The following diagram shows the construction of slur/doubling patches.

Ill. 3.56:
Slur/doubling
patch
(magnified)

61

3.19.7.6 Plate Copy Control Elements

Plate copy control elements are needed only if film is produced that needs to be copied to a plate. Typical plate copy control elements are made up from micro lines under different angles as well as highlight raster patches to make sure that also highlight raster dots are properly copied to the plate. The raster elements exist both as positive as well as negative control elements.

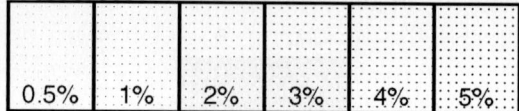

0.5%	1%	2%	3%	4%	5%

3.20 Press Measurement

In order to ensure consistent quality over time, measurement of critical values in the print process are necessary. Two major kinds of measurements are taken to support quality, either density measurements of the ink coating on paper, or colorimetric measurements to check that the printed color is the expected color.

3.20.1 Densitometric Measurement

In order to achieve good and consistent quality of print it is necessary to maintain uniform conditions for the press run. One important step to achieve such uniform conditions is to ensure consistent thickness of the ink coating on paper. This can be measured by using densitometric measuring devices. The density of the full tone control elements is typically used for that purpose. It is important to maintain not only consistency between different prints (e.g. prints taken at different times during a print run) but also consistency across the sheets. There are certain influences of the overall ink distribution across the sheet that require attention.

Every quality press shop maintains certain standards regarding densitometric measurement, which enables a standardized transfer curve for dot gain.

Densitometric measurements control the consistent coating thickness of one ink. In order to include the interaction between different inks, it is necessary to base quality on colorimetric measurement.

3.20.2 Colorimetric Measurement

Colorimetric measurement is typically based on measuring with spectral colorimeters. These instruments either use appropriate filters that approximate the human color response functions, or the measured light is dispersed into its spectral components. The figure below illustrates the operational principles of such a spectral photometer; these instruments today measure in 5 or 10 nanometer intervals across the visible spectrum of light. The instrument type with filters does the integration into X, Y, and Z values by the filter used; in the second type the spectral distribution is measured and the integral values are computed using appropriate weighting functions to generate the basic X, Y, and Z values. From these values the appropriate system values (in most cases CIE L*a*b*) are computed.

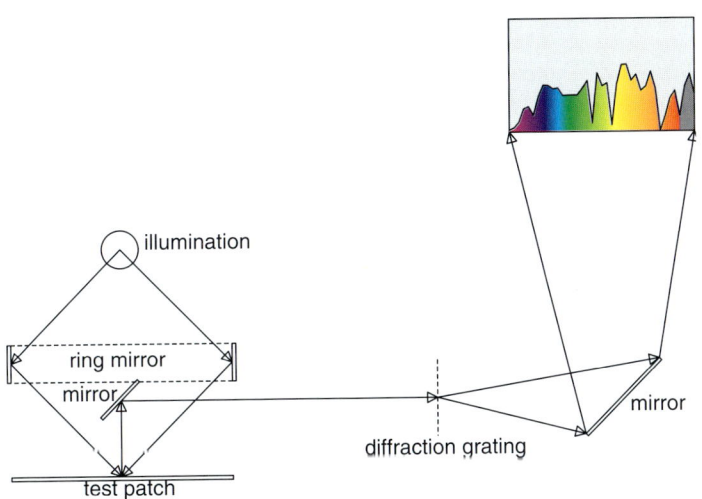

Ill. 3.58:
Schematic diagram of spectral photometer

Based on the knowledge of the ink relations at the measured spot and the colorimetric values (automatic) fine tuning measures can be derived that ensure color stability across a press run (or also in a repeat run).

4 Fully Digital Workflow for CIM for Print

Discussion about computer integrated manufacturing (CIM) for print products requires talking about a fully digital workflow. Traditional methods like cut and paste on a light table are no longer appropriate for CIM for print. This means that the complete process chain must be digital, from the creative systems to the final bound and trimmed brochure. The following steps are involved in a digital process chain for workflow. In all these steps we also discuss the kind of formats that are used as import or as export (where applicable).

Some of the technical issues, especially those related to color reproduction, have been discussed already in the previous chapter. Here we want to put everything into the perspective of the whole production process involved.

Here is just the list of the steps to be discussed, and each of the steps will be discussed separately.

- Creative Systems
- Layout Tools
- Imposition Tools
- OPI Servers
- RIPs
- Trapping
- Image Setters
- Plate Copiers
- Presses
- Finishing Equipment
  - Folding Machines
  - Cutting Machines

    – Collecting Machines
    – Binding Machines
    – Three-Side Cutting Machines

It should be mentioned already here that we will deal with these topics in different depth; for example RIPs will be dealt with in great detail, finishing steps will be covered very briefly only.

The diagram 4.1 on page 67 illustrates the major process steps for a print production and their relation.

4.1 Creative Systems

Let us start out at the creative systems. These include tools for the different kinds of data that need to be created and handled. The following are the important data types to be dealt with: text, graphics, and images. Typical representatives of this group are e.g. Adobe Photoshop or Illustrator or Macromedia Freehand.

4.1.1 Text: Text Editors, Typesetting Systems

Text editors are mainly concerned about the (textual) content. Typesetting is concerned about the visual appearance aspects of type on pages, like typography, the influence of the selected type, its legibility, or attributes of type (like kerning, size, etc.). There are many tools available in this area, and we will not discuss these issues in greater detail.

4.1.2 Graphics: Graphics Editors

Graphics editors are tools that handle drawings and illustrations. Functional elements include:
- constructing drawings,
- to modify them,
- to group elements to complex components that are handled as single objects,
- to store,
- to import, and to export.

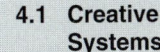

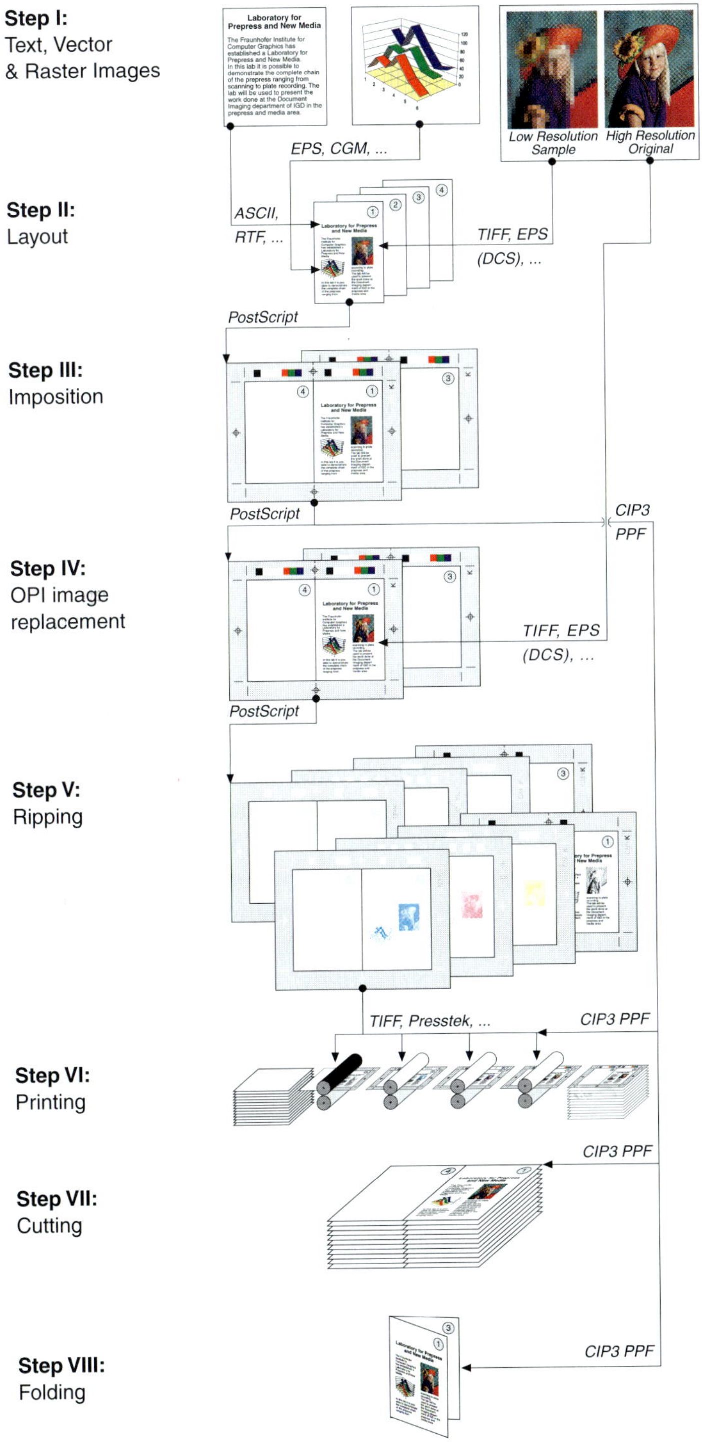

Step I:
Text, Vector & Raster Images
Laboratory for Prepress and New Media
EPS, CGM, ...
Low Resolution Sample
High Resolution Original
Step II:
Layout
ASCII,
RTF, ...
TIFF, EPS
(DCS), ...
PostScript
Step III:
Imposition
PostScript
CIP3
PPF
Step IV:
OPI image replacement
TIFF, EPS
(DCS), ...
PostScript
Step V:
Ripping
TIFF, Presstek, ...
CIP3 PPF
Step VI:
Printing
CIP3 PPF
Step VII:
Cutting
CIP3 PPF
Step VIII:
Folding

Ill. 4.1:
Print production workflow diagram

4.1.3 Images: Scanners, Image Editors

The creation of images can be either done by scanners, if a photograph is the starting point, or by a rendering program that uses a scene (2D or 3D) to generate an image. Besides creation, image editors can be used to modify (scanned or computed) images. All these tools need the possibility to deal with external representations, storage in files, in numerous formats.

At this stage in addition to the high resolution original image a low resolution version is created for later use for layout purposes together with the Open Prepress Interface (OPI).

4.1.4 External Representations (Import/Export): PS, EPS, AI, TIFF, OPI

The number of formats used in creative systems, either as native to the system, or as exchange format or exchange methods between process steps, are quite numerous. The more important ones: PostScript (PS), Encapsulated PostScript (EPS), Adobe Illustrator Format (AI), Tagged Image File Format (TIFF), or the Open Prepress Interface (OPI). The latter would be typically an EPS image file containing (OPI) PostScript comments referencing the associated high resolution image file; this is especially important for performance reasons. More information about the different formats like PostScript or TIFF can be found in chapter 5 *(Formats)*. More details about OPI are discussed in chapter 4.4 *(OPI Servers)*.

4.2 Layout Tools

Layout tools are the means to position content of various types (text, graphics, images) on a page, to shape the visible impression of the page. Page layout is an important aspect of the artistic behavior of a print product. It gives the "look and feel" of a print product, and it may give the distinguishable visual properties of a product. Typical representatives of this group are Adobe PageMaker or Quark XPress.

4.2.1 Import to Layout

The import formats or methods into this step are the same as the output formats of the previously discussed (creative) systems, EPS, AI, TIFF, including the use of OPI.

4.2.2 Export from Layout

The export from page layout systems typically is done in PS, and EPS, also using OPI. Functionally, the output of the layout system are fully formatted pages.

4.3 Imposition Tools

The next step in the chain is to map pages to sheets which later shall be printed (e.g. on an offset press). This process is called imposition. Typical representatives of this group are Adobe PressWise, Farrukh Systems Imposition Publisher, Linotype-Hell Signastation, ScenicSoft Preps, or Ultimate Technographics Impostrip.

The result of placing and orienting pages on a sheet is called a signature. The generation of a signature also has to take into account how the final product is going to be produced. Here an important aspect is how the sheets have to be folded, how they have to be cut, how the pieces have to be collected, and how they have to be bound. This information influences the process of imposing the pages on the sheets.

Under some (simple) circumstances this step is already achieved with the use of page layout systems. This is typically the case when no more than two pages are printed on each side of a sheet. In case of only one page per side of a sheet this step is not necessary. Imposition is a very tedious process that has many parameters that need justification.

We will demonstrate the functionality of imposition with some simple examples. Typical print products contain more than single pages. Let us assume a simple product be an eight page A4 brochure being produced on an offset press which can generate oversize A2 format, i.e. four pages per sheet side. Let us assume the brochure has been produced with some crea-

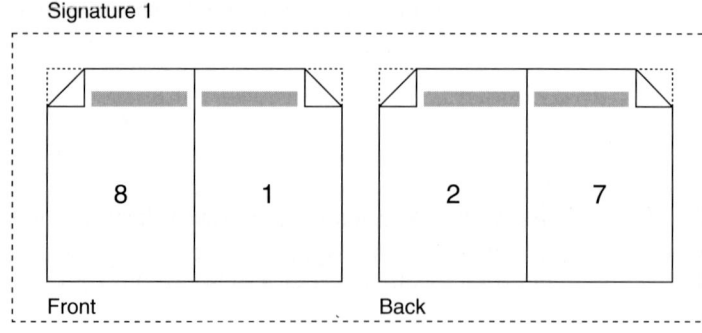

*Ill. 4.2:
Imposition
example resulting
from A2 oversize
press use*

Signature 1

Signature 2

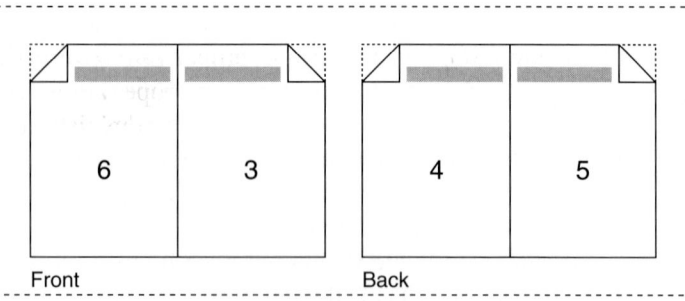

*Ill. 4.3:
Imposition
example resulting
from A3 oversize
press use*

tive tools; their output is a sequence of 8 pages in PostScript.
In order to be able to print this brochure on A2 oversize sheets
it is necessary to place the right pages in the right place on

film, and in turn then on the front and back sides of the sheets. Placement and orientation of respective pages has to be properly chosen.

As an example let us show the imposition result of the brochure mentioned above, using an A2 oversize sheet fed press (as mentioned above), and in another case using an A3 oversize press. (4.2 and 4.3 on page 70)

In addition to correct functionality there may be additional constraints to be observed to get a quality product as a result.

For example, if in our brochure we have a crossover, i.e. one article goes over two pages side by side, and there is a heading (or even an image) placed across the two pages; of course these should align well across the two pages. In production it is always easier to place such an article in the center–fold of the brochure, where the two pages are side by side on one sheet anyway. Any other placement would make the accuracy requirements on the finishing of the product (collecting, aligning, cutting) much higher (however still possible).

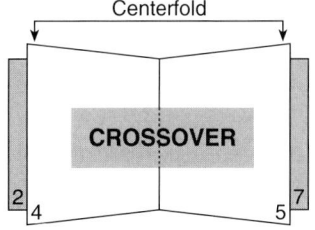
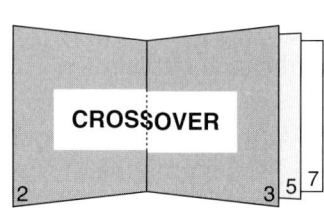

Ill. 4.4: Page crossing alignment problems

Other pieces of information are included besides placing the net page information on the sheet. These include objects like color control strips (for checking proper ink supplies on the press), registration marks (to check the proper alignment of separations), cut and fold marks, as well as administrative information.

4.3.1 Input to Imposition

The import formats or methods into this step are the same as the output formats of the previously discussed systems containing full page information, namely PS and EPS, possibly using OPI.

4.3.2 Output from Imposition

The export of page layout systems typically is PS, using OPI. Functionally, the output of the imposition system is ready-to-print sheets with still low resolution images in place. The imposition step may also supply information which is needed further down the process pipeline in form of CIP3 Print Production Format (PPF) files. At least all the information about cutting and folding is available at this stage. A more detailed discussion can be found in chapter 8.4 *(The CIP3 Print Production Format)* later in this book.

4.4 OPI Servers

One of the practical solutions to the integration problem (at least from the point of view of the user) is the Open Prepress Interface (OPI) [AL93]. The OPI concept takes into account that high resolution image data is large as compared to text, and that layout software typically uses low resolution versions of the high resolution data for preview and placement of images on pages.

OPI was introduced by Aldus Corporation as a set of PostScript comments that are used together with the low resolution images; the comments provide information about the location of the related high resolution files together with placing, cropping, scaling, rotation, and color information. Since this information is syntactically hidden in PostScript comments, they can be processed without interpretation of the PostScript code. The following code excerpt shows an example of OPI comments as they would appear in a file with OPI references.

```
%ALDImageFileName: ATS:Clients:Covaldi.eps
%ALDImageDimensions: 197 173
%ALDImageCropRect: 0 0 171 160
%ALDImageCropFixed: 0 0 171.079 159.74
%ALDImagePosition: 19 18 19 178 190 178 190 18
%ALDImageResolution: 72 72
%ALDImageType: 4 8
%%BeginObject: image
```

```
    % ...at this position the OPI servers
    % includes the high resolution image data...
  %%EndObject
```

This allows application programs to substitute the low resolution data with the high resolution data before processing it in the RIP. These programs are typically called OPI servers, because they provide image replacement as a service before producing the final output on film, plate, or printing machine.

Different OPI servers provide different functionality. Some servers provide format conversion, some only PostScript integration. The effect of this type of PostScript integration however is the explosion of PostScript file sizes, especially when PostScript code for images does not take advantage of compression mechanisms provided by Level 2 RIPs.

Typical representatives of this group are Helios EtherShare, Luminous Color Central, or ScenicSoft Printdesk.

4.4.1 Import

The input formats to the OPI server are the same as the output formats of the previously discussed systems describing full sheet information, namely PS, containing OPI references.

4.4.2 Export

The export of OPI servers is PS, with all OPI references resolved and the high resolution images included in the PS file. The format of the high resolution image files referenced is mostly TIFF. The OPI server converts the TIFF image to proper PostScript image code. Functionally the output of the OPI server is ready to print sheets including all information that can directly go to the PostScript RIP.

4.5 RIPs

The term raster image processor or RIP is used with various meanings. The term either is used for the processor that converts PostScript or more general a page description language (PDL) to the device's representation – the output is directly usable for the destination device – or for the processor that converts a continuous tone image into a screened representation to be used with a device that can only handle one bit color depth per process color. The first usage of RIP is a superset of the second.

Typical representatives of RIPs are Adobe's Configurable PostScript Interpreter (CPSI), or Harlequin's ScriptWorks.

In the following sections we will discuss some aspects related to RIPs that are of great significance to their proper operation, the selection from various alternatives, and the installation environment.

4.5.1 Structure and Architecture of RIPs

In the past RIPs have often been considered as Black box items. Often they were built as special hardware that allowed little external control. Today most RIPs are pure software that run on standard platforms. Actually, much more than the pure rastering process is embedded in a RIP. RIP software can be a complex network of distinct processes sometimes implemented in a single program, but also often as a set of cooperating processes at the operating system level.

Server architectures also play an important role in the context of RIPs. Servers are used especially where special resources are needed, e.g. for supplying fonts, accessing high resolution image data, or applying high quality screening algorithms based on large amounts of precomputed information.

The complex processes forming a RIP also include a significant amount of uncertainty whether a delivered file can be properly processed by the RIP. This fact prepared a market for so-called preflight systems – systems that perform certain plausibility checks on the input files to find missing or wrong information before costly processing happens further down stream.

The architecture of RIPs must also allow proper exception handling in case that error conditions arise. Most RIPs are part of a larger system connected by a certain workflow that does not allow stopping of a process in the middle of the pipeline.

This requires special attention for the setup of the system. Other technical questions play a critical role, e.g. how to configure a network environment for a RIP such that the capacity of the RIP is not limited by inefficient network structures.

A typical RIP contains (at least conceptually) the following components:

- language parser
- scan converter
- font manager
- font rasterizer
- color manager
- image scaler
- display list processor
- rasterizing processor (screening)

These are just the more obvious components; the smooth integration and interoperation of these components makes a high performance RIP possible.

4.5.2 Processing Graphical Primitives in a RIP

The input data for a RIP contain different kinds of data types. For our consideration here we will distinguish between graphics (described analytically), type, and raster images. Post-Script provides constructs for all three types of page elements.

Graphics primitives are described by a geometric description of their outline; it may contain curves and lines as boundary elements. This geometric outline, in PostScript called path, is bound to a color at the execution of the operator to fill that outline. Even objects generated with a `stroke` operation are internally at least conceptually handled like outlines to be filled. The process of scan conversion determines the set of device pixels covered by the describing geometry. These device pixels are associated with the appropriate color.

Typefaces are defined by the set of glyphs belonging to that typeface. However, glyphs are not just ordinary graphics. The process of scan conversion for glyph outlines differs significantly from the scan conversion for graphics primitives. The scan conversion process for glyphs sets less pixels than the scan conversion for graphics primitives. Besides this difference in the rules for which pixels to set, there is normally a

75

hinting process associated with fonts that ensures that certain properties of fonts are retained, especially in small font sizes (in number of pixels). For example it is necessary to make special provisions for serifs not to disappear at sizes smaller than 10 pixels. In addition the result of this step for a glyph is stored in a cache – the so-called font cache – to avoid repeated execution of the relatively compute intensive scan conversion process.

As opposed to graphics primitives, glyphs and their position are always mapped to the device raster. When drawing glyphs to the page buffer the shape is taken from the cache and painted in one color to the page.

Ill. 4.5:
Conventions for
scan conversion of
type and graphics

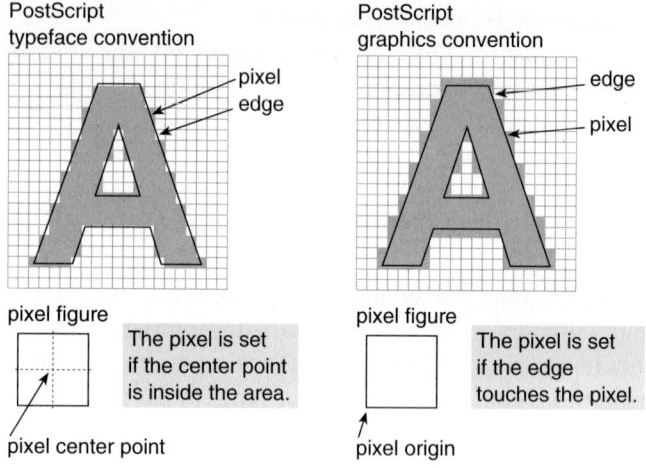

Ill. 4.6:
Times-Roman
typeface output
from original
Type 1 font and
from Type 3 font

Times-Roman 8pt, as the original Type1 font

Type1 18pt

Times-Roman 8pt, converted to Type3

Type3 18pt

The scan conversion for graphics sets all pixels that are touched by the edge of the borderline in addition to the pixels completely inside the borderline. For scan conversion of typefaces the pixels are set if the center of the pixel is inside the boundary.

The top example shows text that was scan converted by a font engine; the bottom example was converted to geometry and scan converted by the standard graphics engine. The following illustration shows these results in a magnification that allows to view pixels. It takes the text portion *Type* from these examples, and to the right shows the pixel difference of the two types of rendering the underlying geometry.

Type
Type

Ill. 4.7:
Magnification of
illustration 4.6
with pixel
difference
between Type 1
and Type 3

A raster image represents a function $f(i,j)$ of color values by a two-dimensional array. This function can be interpreted with two different rules, either sampling or interpolation.

Sampling used for interpretation of an image determines the color value of a device pixel by inverse mapping of the device coordinates of the pixel to the image array. If the image array cell is large compared to the device resolution (which is normally the case) this leads to the effect that the pixels in the area of the cell all are set to the same color value (i.e. the value of the cell).

Interpolation used for interpretation of an image assumes the color values in a cell of the image array is correct for the center of the cell. Color values that map to a position different from the center of the cell are evaluated by a bilinear interpolation of the participating four color values of the neighboring pixels.

The reason for using interpolation can be understood if looking at the patterns created when image pixels and screen cells interfere. The following diagrams (4.8–4.11 on pages 78 and 79) show the differences of sampling and interpolation on the screened results of an image.

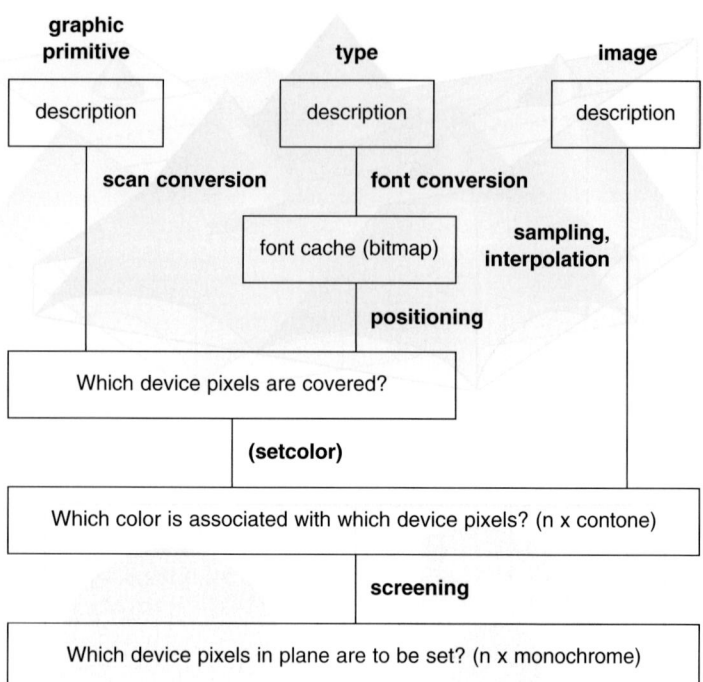

Diagram showing the graphics primitive, text and image processing pipeline

In a RIP, temporal ordering of the execution of the various steps generating output can vary greatly. This is demonstrated by looking at what happens at the time when an outline is filled, (`fill` time) and what happens, when a page is to be painted to the device (`showpage` time). The RIP is assumed to drive an image setter. This is a simplified look but it gives a good insight in some of the architectural questions associated. We can distinguish three major architectures. We describe the chain in terms of the following (at least conceptually present) process steps:

- [1] language interpretation
- [2] building the data structure
- [3] generating four separations of 8 bits each
- [4] generating four separations of 1 bit each
- [5] sending bitmap to the device.

Looking at the execution of the steps only the time, when a `fill` is executed, and the time when the page is painted to the device, is considered.

The figure 4.13 on page 81 gives a schematic overview of the three alternatives

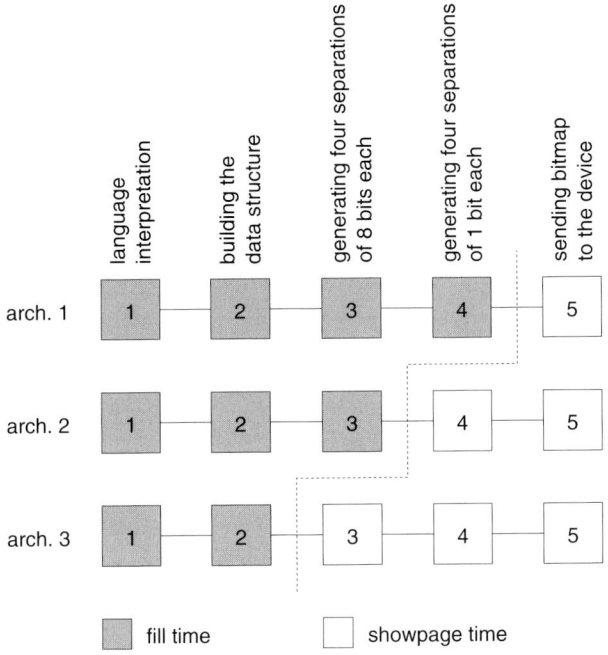

Ill. 4.13:
Alternative RIP
architectures

One architecture does everything except step 5 at the time the fill is executed. Only the sending of the bitmap to the device remains to be done at showpage time.

The second architecture executes steps 1 through 3 at `fill` time. The screening process is delayed until the page is written to the device (steps 4 and 5).

The third alternative architecture just executes steps 1 and 2; it generates a high level data structure often called a display list; the scan conversion and screening are delayed until the page needs to be output after a showpage operator has been encountered.

These distinct architectural structures need to be known and taken into consideration to understand the resulting effects when discussing the further topics regarding RIPs.

4.5.3 RIPs and Image Setter Device Effects

Input to RIPs specifies anticipated output results. This, however, requires that RIPs adapt to the specific behavior and deficiencies of the respective output device. In the case of image

Ill. 4.14:
Area of pixel
cluster with
laser diameter
of √2 times the
device resolution
(shown as
square grid)

setters there are several problems that need special attention in order to produce good quality results.

An idealized device model of an image setter (and many other devices) partitions the device space in an array of rectangular (usually square) cells. With this concept it is relatively easy to produce a certain tonal value: the tonal value (between zero and one) can be computed from the ratio of pixels set in a raster cell to all pixels in the cell. So both for AM and FM screens this seems to be quite simple: just make sure that the right number of pixels is set per cell area. The details of screening already have been discussed.

In practice, device pixels are not square at all – typically they are round or even elliptic. Taking this into consideration the RIP needs to compensate for effects produced by this fact. Ideally, the diameter of the elementary device pixel is $d=a*\sqrt{2}$, if a is the length of the square of a device cell, because this is the smallest round shape that can fully cover the whole output space. This means that the area covered by a set of device pixels is generally larger than indicated by the number of pixels set. This effect is similar to dot gain and is more significant in

FM screens than in AM screens; middle tonal range of FM screens produce the most noticeable effect. Obviously the tonal effects that are generated are heavily dependent on the screen. To compensate for these effects a linearizing transfer function is applied to the tonal value before calculating the screen, and this function has to be tuned to the screen used. The figure 4.14 on page 82 illustrates the area increase with ideal laser dot size showing some simple pixel configurations.

Devices are not always ideal with respect to the size of round pixels. There are devices that produce much larger sized pixels. The following example (4.15) shows results from a device that utilizes larger device pixels with the same pixel configurations as in the previous example. The partitioning of the device space into curve bounded pieces can be used to calculate theoretical transfer functions that map quite nicely to practical experience.

In addition to the effects caused by the size of round pixels the effects of dot gain have to be compensated for. As pointed out earlier they are also dependent on the type of screen used.

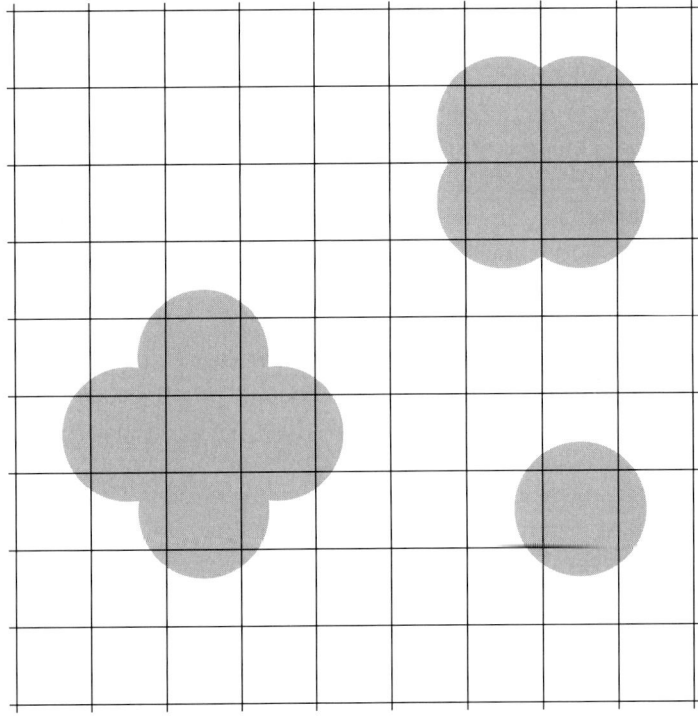

Ill. 4.15:
Area of pixel
cluster with laser
diameter of 1.75
times the device
resolution (shown
as square grid)

These effects can be compensated in a combined transfer function so there is only one function necessary to apply.

4.5.4 Performance of RIPs

Performance is vastly different for different RIPs and different jobs. Benchmarking of RIPs is extremely difficult. Some RIPs do not exhibit a reproducible behavior as they permanently use the idle time of the RIP to prepare results that are likely to be needed in the near future. This is especially true for font rendering and caching.

Different RIPs also show common behavior, e.g. in the amount of time used for screening a page for halftone output. Typical figures are a distribution of 80 to 90% for screening compared to 10 to 20% for the rest of the PostScript interpretation. These are typical figures for a software RIP producing 1270 to 2540 dpi (500 to 1000 device pixels per cm) output for an image setter.

General timing differences between different RIPs under identical environment conditions with identical jobs can vary easily by a factor of 20. This factor between RIPs can well be significantly different for different jobs. Some RIPs do better with image output than other, or some RIPs have critical memory consumption behavior.

The configuration of a RIP environment can be of critical importance for the performance of a RIP. The right configuration can well make a difference of a factor of two to three (or even more) in speed. This problem will be discussed in more detail in the section 4.5.6 *(Environments for RIPs)*.

There are numerous benchmarks for PostScript RIPs. It is however critical to understand what exactly these benchmarks measure. In many cases (e.g. the well-known Seybold test suite) what is really measured is not the RIP speed, but a speed imposed by the whole environment including the RIP. This means that these figures typically do not say very much about the speed of the RIP itself. One can argue, of course, that in all practical conditions a RIP is used in a certain environment, and if this environment is used also for the benchmark then the results should be useful for the situation. Still our experience shows that effort spent in a little tuning makes significant differences in the resulting performance improvement, and the tuning for different RIPs may need different measures to get the best results.

4.5.5 Functionality of RIPs

The functionality of RIPs also is an area of great diversity. Whereas hardware RIPs were predominant in high performance solutions until a few years ago, today most RIPs are software, sometimes utilizing a special purpose hardware for time critical functions. Here special hardware support for screening is the most obvious candidate as pure software screening at high resolutions takes, as mentioned earlier, in the order of 80% of the total RIP time.

A classification for RIPs could be established based on a few important properties: PostScript Level 1 or Level 2 (see sections 5.1.1 and 5.1.2 for discussions of Level 1 or 2), halftone or continuous tone output, size limitations (maximum format), high quality screening (AM, FM) for halftone output available, file system access (for direct execution of OPI references without OPI server) available, font library and quality font rendering, and benchmark figures. These properties allow us to distinguish good RIP implementations from mediocre or even bad ones.

4.5.6 Environments for RIPs

Critical to the smooth functioning of a RIP is the appropriate setup of the environment. Hardware and system software configuration plays an important role here. For today's high quality high resolution RIPs it is important to balance input/output (I/O) to disk between the access to possibly extremely large PostScript files and the intermediate data I/O necessary to hold large display lists before output of the final result can take place. Moving intermediate data to a separate disk path may result in extremely large performance increase, whereas conflicting I/O on one disk path may degrade an otherwise powerful system to a bad performing system.

85

Not only hardware and system tuning are required; the RIP resources have to be set up to allow smooth operation. Typical setup informations for the RIP include

- font resources (i.e. which fonts are available, and where they can be found).
- screening information (some systems allow the overriding of screen settings in the PostScript files to ensure consistent quality output in a production environment),
- information about input and output queues,
- intermediate storage allocation,
- monitoring procedures,
- possible inclusion of OPI server information, etc.

All these RIP configuration details have to be initialized according to the requirements of the RIP job processing at the shop. Overall performance and throughput considerations are dominating the decisions for the RIP setup.

The previous information should also be regarded as a source of bias for different RIP benchmark results. In the end it is necessary for a RIP to show good performance in real operation rather than in benchmark results. According to our experience it is necessary for every RIP to do proper tuning and configuration in order to get the best possible results with that single RIP. It is not sufficient to install a RIP and measure; results achieved by that kind of testing are not valid for a statement about a specific RIPs performance compared to another RIPs performance. Different RIPs are distinct in their reaction to changes in configuration, and this is due to differences in the internal architecture. It is, however, safe to say that it is always possible to achieve significant improvements in performance if the necessary research is put into the tuning.

4.5.7 Controlling the RIP Process

Production speed is of critical importance in real life businesses. Therefore, controlling the RIP process is a sensitive issue. In some environments we may be able to control everything from customized windows interactively, but in many shops everything has to run on its own, without necessarily interacting with every job. This is prerequisite for an automated production chain of which a PostScript RIP is part of.

What needs to be controlled? Critical areas are those that affect the quality of the result and the use of resources. The following two sections deal with these important factors.

4.5.8 Quality Issues for RIPs

Besides the correctness of the interpretation the most critical issue regarding quality probably is the rastering or screening process. Here we have to distinguish between different alternatives:

- Use that which is specified in the PostScript file. This leaves us with large variations in quality across jobs. One job may produce nicely, another job may produce mediocre (color) quality. Normally a PostScript driver does not know enough of the RIP environment to make intelligent decisions here.
- To override whatever is specified in the PostScript file with regard to screening. This allows to produce consistent quality across jobs. The consistent quality can be good quality if the effort was undertaken to run an extensive set of tests to find the best parameter settings for the specific output process. The parameters selected here are typically the angles, the frequencies, and the dot shape of the screen for the respective separation. This however leaves us with the restriction that specific effects that are generated using specialized spot functions for screening can not be reproduced – unless in such cases it is deviated from the standard process.

Whichever way one chooses it is still important to have a quality control installed. Even in a fully computer integrated production environment, visual control before results leave the shop is always a good idea, and it can ensure the satisfaction of customers.

4.5.9 Resources for the RIP

Resources required by the RIP are both hardware and software resources. Some hardware resources are fixed by the decision to use a certain system for the RIP; an example here is the processor type. Critical hardware and software resources are determined by the integration of the RIP in the workflow via a network.

Resources are quite manifold, and we will discuss here only the more critical ones. So, for example, the assignment of main memory and disk space for the operation of the RIP can make extreme differences in the behavior of the RIP. Typical RIP-related resources are the installed fonts and precomputed screening information. We have discussed the problem of having the correct fonts available for the RIP earlier. The precomputed screening tables are a means for increasing the performance of the RIP for preselected screen frequency, angle, and dot shape combinations. This is typically done for those combinations which have proven good quality in tests before.

4.5.10 RIPs as Part of a Workflow Solution

RIPs are core components in the workflow from creative systems to quality print. The normal workflow however contains a lot more components than just RIPs.

Let us briefly summarize the workflow discussion so far, and put the role of the RIP in perspective:

- Creative systems typically generate from their native internal format PostScript files that contain references to high resolution image data. Based on these PostScript files often an imposition step is applied.
- The next step often is the integration of the high resolution images into the PostScript files using an OPI server. This processing step generates very big files.

An important question remaining is whether the fonts have been included. If they are not included – and some packages do not include fonts - it is necessary to download the required fonts to the RIP or use some tool to integrate the fonts into the file. When this stage is reached the time is come for ripping. Often RIP errors occur due to some subtle error in previous processing stages. To avoid expensive RIP runs (film or plate consumed, and time consumed) often so-called preflight systems are used to check the consistency of files, e.g. to check whether all required fonts are supplied, whether all low resolution images have been replaced by their high resolution equivalents, or whether correct PostScript code has been supplied. Some preflight systems do a full PostScript interpretation. Preflight systems can reduce RIP failures significantly, but – due to the complexity of PostScript and RIPs – they cannot eliminate them completely. Here follows the RIP process.

Taking the workflow beyond prepress to press, information about the ink consumption and resulting preset values for the press can be derived from the raster image data. The state-of-the-art way of presetting presses is an expensive task that requires additional manual steps for the pressman. In order to calculate correct area coverage in the RIP (or from the RIP generated bit image) that can be used to derive press settings, it is necessary to know device specific parameters like device resolution and the diameter of the effective laser beam. With this information available presses can be brought up to production running much faster, and less garbage is produced during setup of the press. Solutions for this type of information exchange have been produced by the Fraunhofer Institute for Computer Graphics in cooperation with Heidelberger Druckmaschinen AG and a large group of prominent vendors from prepress, press and postpress. The name of this information exchange solution is CIP3 Print Production Format; CIP3 stands for *International Cooperation for the Integration of Prepress, Press, and Postpress.* More information about this development is given in chapter 8.4 *(The CIP3 Print Production Format (PPF)).*

4.5.11 Input into and Output from RIPs

The input into RIPs today is – in practice – only PS.

The output from RIPs is first of all device resolution bitmaps that contain all information to generate film or plates on an image setter. The CIP3 PPF files can also be generated during this step (for later ink calculations)

4.6 Trapping

Overprinting multiple separations requires exact alignment of the different separations to avoid unpleasant effects. For example, if one places a Black text on top of a Magenta area, a slight misalignment of the two involved separations (M and K) results in white shadow regions around the text.

The result is not like painting the complete area with Magenta, and then painting the text on top of it. If the output format is PostScript (as is normally the case), the PostScript imaging model clears all separations under the text and sets the Black separation in the appropriate areas filled by the text.

This method would require exact (in sub pixel scale) alignment of the separations to avoid paper White to shine through at the edges of the text glyphs. Exact alignment however is difficult due to the fact that paper is not static; it typically streches due to ink deposited on its surface,and this paper movement changes alignment of separations beyond control.

Trapping is a method that allows to relax the strict requirement on sub pixel alignment. In our example, the Magenta separation would extend under the text glyphs; the size of extension that is necessary depends on the process to be used for printing. Trapping does not eliminate completely all effects of misalignment, but it reduces the visibility of these effects significantly.

Tools today allow trapping typically in two places, either in the PostScript generating application, or in the RIP.

4.6.1 Trapping in Applications

One is done on the side of creative tools, e.g. by modifying the paths of objects to extend or shrink in certain regions, depending on the separation and on the colors of overlapping objects involved. This method does not take into account the particular effects produced by the printing method.

4.6.2 Trapping in the RIP

Alternatively one can apply trapping at the level of the generated bitmap for output to the device; here combinatorial considerations can be used to detect the critical areas and reduce the undesired effects generated by slight misalignment of the separations. With this method even complex situations – involving many overlapping objects – can be coped with.

As the implementation of an optimal trapping is process dependent, best results can be achieved if trapping is not executed in the generating application, but at the device dependent bitmap.

The process of trapping at the bitmap level is illustrated in the following figures.

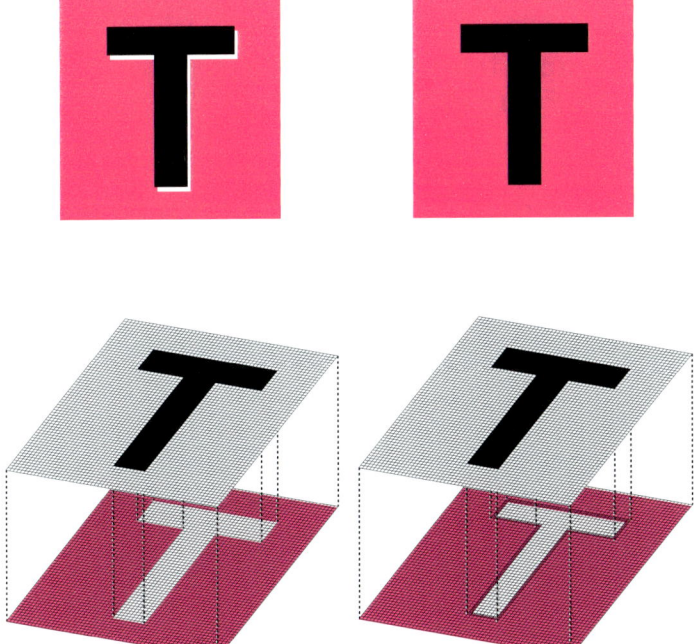

Ill. 4.16:
Trapping at the
bitmap level:
left half no
trapping, right
half trapping
applied

4.7 Image Setters

Image setters are high precision devices that produce device pixels on a light sensitive material. They apply a laser beam directing the light to the spot to be imaged through an arrangement of polygonal mirrors. Depending on the material to be imaged, laser light of different wavelengths is used. After the imaging process normally a development process is necessary to fix and stabilize the image on the material.

4.7.1 Input into the Image Setters

The input into an image setter is the bitmap of the sheet to be imaged coming out of a RIP. The information on that bitmap is typically a bi-level image, i.e. pixels can either be set (exposed by the laser beam) or left untouched. Depending on the process and material positive or negative images are required (either pixels set are imaged, or pixel not set are imaged). The process of screening has already been applied in the RIP.

4.7.2 Output from Image Setters on Film or Plate

The output of an image setter is typically either film or plate; often directly connected to the image setter is the development process; it may be a normal chemical development and fixation of film, or it may be a thermal development process as with certain printing plates in use today. The final result is a carrier material containing the image of the sheet to be printed by one of the process colors. Following a schematic diagram of a plate setter with associated development unit and plate stacker is shown.

Ill. 4.17:
Schematic
diagram of a
plate setter with
development unit

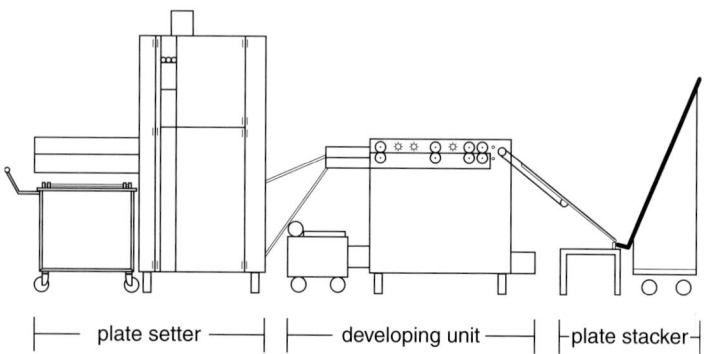

├──── plate setter ────┤ ├──── developing unit ────┤ ├─plate stacker─┤

4.8 Plate Copiers

In case of film output from the image setter a further step is required before going to the press. The film is copied to the plate using an analog copying process. This process is very critical especially for small details on film; typical problem areas are films generated with FM screening which contain small dots. In this case it is especially important that the exact exposure time for the plate is observed within tight limits.

4.8.1 Plate Copy Input and Output

The input into this process step is always film; in more conventional processes this stage may be preceded by a step doing cut and paste of film to generate a full sheet size image. In a fully digital process however this does no longer happen.

The output of this phase are printing plates; depending on the type of material used different development steps are required.

4.9 Presses

There are many different press types employing different printing processes like flexographic printing, gravure printing, silk screen printing, or offset printing. We will limit our discussion to the often used offset printing process. Offset printing is an indirect printing process which, different form other processes, prints from a flat plate surface. In letterpress printing for example only those areas print that are elevated; in gravure printing ink is transferred in places where the drum has been engraved, i.e. is lower, and therefore contains ink. Offset uses the fact that oil and water normally do not mix; the surface of the plate is conditioned to attract ink in those places where an image shall be transferred. We will explain this process in more detail now.

An offset press consists of one or multiple print units. Each print unit has five major components:
- the plate cylinder, associated with
- an inking unit, and
- a damping unit,
- a rubber sheet cylinder, and
- an impression cylinder.

Two schematic diagrams (4.18 and 4.19) of a print unit are shown on pages 94 and 95, the second being a bit more detailed than the first.

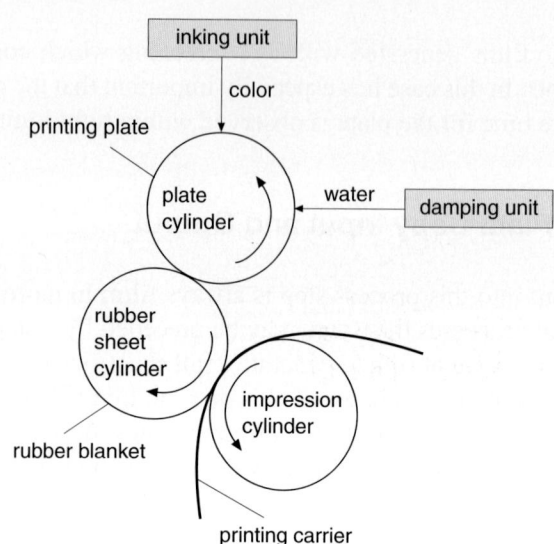

Ill. 4.18: Schematic diagram of an offset print unit

The printing plate is mounted on the plate cylinder. The surface of the printing plate is oleophilic or hydrophilic, depending on whether ink shall be transferred or not.

The damping unit applies water to the surface of the plate via the damping duct and vibrator rollers, the damping distributor and the damper roller.

The inking unit contains the ink duct (which contains the ink supply). Ink is transferred via the ink duct roller and the ink vibrator roller (to control the amount of ink supplied). The many further rollers are used to reduce the thickness of the applied ink coating to the required measure of roughly one micron. The plate inkers finally transfer the ink to the plate.

The inking unit is horzontally divided into multiple ink zones each of which is controlled by an ink key. Each ink key supplies the required amount of ink for the specific zone that ink key controls. Depending on the amount of ink needed for the subject printed in that zone, the ink key supplies more or less ink. The accurate control of the ink supply is critical in order to be able to maintain consistent thickness of the ink coating finally reaching the paper. There is a lateral distribution from one ink zone to the next that implies ink exchange from highly inked zones to neighboring zones carrying less amount of ink.

The ink image on the plate then is transferred to the rubber blanket mounted on the rubber sheet cylinder. The ink image

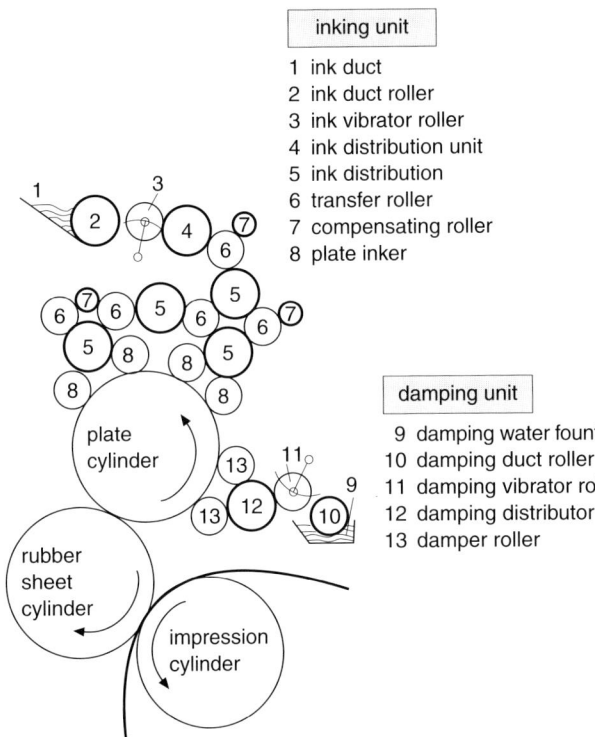

inking unit

1 ink duct
2 ink duct roller
3 ink vibrator roller
4 ink distribution unit
5 ink distribution
6 transfer roller
7 compensating roller
8 plate inker

damping unit

 9 damping water fountain
10 damping duct roller
11 damping vibrator roller
12 damping distributor
13 damper roller

Ill. 4.19:
Detailed
schematic
diagram of an
offset print unit

is in turn applied to the paper that is fed with pressure between the rubber sheet cylinder and the impression cylinder.

4.9.1 Import: Plates, CIP3 PPF

The primary input to the press – actually to the print unit – is
the plate. The plates control where to put ink on the print
substrate.

The amount of ink used for printing is controlled across
the width of the press. So-called ink keys control the amount
of ink supplied in each of several ink zones across the press.
The settings of these ink keys is critical to maintain consistent
quality print on a press. In conventional offset processes these
settings were derived from scanning the plates and computing
the presetting of the ink keys. Today either the CIP3 Print Production Format (PPF) or the PostScript file is used to calculate preset values for the ink keys. In direct-to-press or direct-

to-plate environments this method is preferred, or even re-quired. It produces less waste sheets to print, and puts the press quicker into production.

4.9.2 Output: Print Sheets

The output of a press is always printed sheets of paper (or other substrate). These are typically arranged in stacks. In some cases this may already be the finished product, e.g. if we deal with a poster production. In other cases these stacks have to be further processed in finishing steps.

4.10 Finishing Equipment

Additional finishing steps need to be applied to form a final product like a brochure, a book, or a package. The machines used for that purpose are manifold and different in operation. There is more than one possible sequence of application of finishing equipment. The following is a non-exhaustive list and their respective input and output;

- Folding Machines
  - Input: Print Sheets
  - Output: Folded Sheets
- Cutting Machines
  - Input: (Folded) Sheets
  - Output: Cut Blocks
- Collecting Machines
  - Input: Folded Sheets
  - Output: Collected Sheets
- Binding Machines
  - Input: Collected Sheets
  - Output: Bound Product
- Three-Side Cutting Machines
  - Input: Bound Product
  - Output: Trimmed (Finished) Product

After finishing packing and palletizing may be applied, or individualizing, e.g. for mail distribution. At this stage the product is ready for the customer.

5 Formats

This chapter deals with various formats found in document applications. We are not discussing internal product formats but rather only formats that can be exchanged, be integrated, or be used as input to a document imaging process. This is merely because there are so many internal formats that will be of use only in very restricted environments. We rather see the multiplicity of interchange formats that is prevalent today. Typically there are different tools for different kinds of tasks, and these different tools tend to generate different output formats. Some of these formats are discussed in the following chapter. The ordering of formats is not to be understood as priorization or value of a format In this chapter we will discuss mainly page description languages, namely PostScript and related formats; TIFF is another format that is covered; besides these two groups, some other formats are dealt with more briefly.

5.1 PostScript

When Adobe came out with its first version of the PostScript Language Manual [AD90] in a ring binder version in 1984, it was not at all clear that this language would make such an impact on the market we are looking at. Interpreters were very slow, devices with PostScript were difficult to find. With the advent of the LaserWriter the success story of PostScript began. In order not to spread too much optimism, it should be noted that the enormous success of PostScript is not necessarily due to the magnificent concepts, or to being the best solution imaginable, but rather to the fact that it provided much more than any other printer format could offer at the time.

PostScript offered possible integration of high quality rendered text, graphics capabilities and the possibility to include images, at first gray scale images only, but with the extension to more color capabilities also color images. So why are we continuing with further formats, if this one is the solution? Some important features for an acceptable general solutions are missing even from the latest version of PostScript, and we will discuss this further later on.

Besides offering page description capabilities, PostScript is a full-fledged programming language, offering features like variables, control constructs, procedures, and files.

5.1.1 Level 1

The term Level 1 in relation to PostScript was introduced only after a Level 2 PostScript had been described; it served as the label for the original PostScript as brought out in the mid-eighties. It included however some color extensions mentioned above, which were introduced, because prepress applications could not use PostScript without such extensions; part of these extensions were operators supporting the CMYK color space.

The merits of PostScript lay especially in the handling of text. The concept of intelligent font scaling was introduced in a convincing manner. Fonts were no longer fixed sizes, 10 pt. or 12 pt., but were available in any scalable size. This however caused a lot of typographers to view PostScript with negative attitude, because they argued (correctly) that fonts are not really geometrically scalable. The technique of using hints for the rasterization of fonts was developed further, and size-dependent processing of fonts became possible; today this technology is even accepted by most typographers.

The format used for font descriptions in PostScript, the so-called Adobe Type 1 format [AD90a], is an encrypted format which was kept secret until 1990. This fact kept independent font suppliers from offering Type 1 fonts. The format was opened only after Apple and Microsoft announced the opening up of their TrueType format [MI90].

Over the years thousands of fonts have been developed by Adobe and others. Type 1 is the format with by far the largest number of high quality fonts available today.

A significant drawback of Level 1 in the area of graphics was the lack of patterns for filling areas. Patterns had to be emulated abusing screening to provide fill patterns. In general a significant problem of Level 1 was the rudimentary color support. These deficiencies were finally corrected by introducing PostScript Level 2.

5.1.2 Level 2

With the advent of PostScript Level 2 some problems of Level 1 seem to have disappeared. Level 2 provides patterned fill for graphics, it provides improvements in areas the user does not normally recognize, like better and controllable free storage handling in interpreters, it provides additional simple graphics primitives, it provides a more efficient handling of text, and it offers a complete color concept with device dependent and device independent color.

This last improvement is probably the most important one; it was the prerequisite for being acceptable in the high end printing world. In a time with a high proliferation of color printing devices this is an important step forward towards calibration of color across devices. Level 2 does not provide the solution itself, but it provides the necessary hooks in order to be able to solve the problem.

Another important extension brought about by PostScript Level 2 is the ability to use a variety of built-in compression and decompression methods. This allows for a significant data reduction especially in the area of image data. The filter concept of Level 2 is the technical means providing this compression/decompression facility. This concept was later taken over by the Adobe PDF, the Portable Document Format, used by their Acrobat series of products.

It is also fair to say that Level 2 brought a harmonization of features necessary for devices with certain capabilities that were originally not part of the language definition of PostScript. These extensions make it easier now to address device capabilities like multimple paper trays, ot duplex printing.

5.1.3 Encapsulated PostScript

Another term hit the market after the introduction of PostScript: Encapsulated PostScript (EPS) [AD92c]. First of all EPS, as it is often called, is PostScript, however with some restrictions on the use of certain operators, and some special sequence and comment conventions.

An EPS file contains only one single picture, it starts out with simple comments that give the originator, the resources used (e.g. fonts), the bounding box, and other useful information, and it gives this information in a format that can easily be scanned by application software without interpreting Post-Script. This is important because interpretation of a PostScript file can be an expensive task.

In EPS files operators that induce global and irreversible changes to the PostScript interpreter's internal state are forbidden.

An EPS file may contain a device dependent or device independent low resolution raster image of the content of the file, which the including program may be able to display. So the main intention of EPS is to allow integration of different PostScript objects coming from different applications.

5.1.4 Integrating different PostScript sources

Integration of different PostScript sources is an interesting and sometimes surprising task. Integration of EPS into normal PostScript is less critical, and as we have seen, this is the original intent of EPS. This requires a normal Postscript file (or even another EPS file) into which EPS files can be integrated; the integration works by encapsulating the EPS content between invocations of the save and restore operators to keep the interpreter state unchanged. As long as EPS restrictions are followed closely (and no dirty tricks are used in the EPS file) no problems should occur; however if these restrictions are not observed, very often problems arise that lead to unpleasant surprises, when trying to print. This is also often the case, when trying to integrate multiple normal PostScript files.

A special issue is the availability of document fonts required. This problem can be solved either by making all necessary fonts available (e.g. through downloading the fonts, or including the fonts in the PostScript stream), or by substituting the fonts with alternative fonts approximating glyphs with the same metric and similar design. Since font names are subject to copyright, fonts with identical glyphs and metrics often can be found by correct substitution; so there are about 40 different names in widespread use for the well known Helvetica font family. There are new solutions using the concept of a font finder that provides the necessary font name substitution. An example of such a solution can be found in the results of the EuroRip project discussed later. See 7.3.8 *(Font Server)*.

5.1.5 Color Concepts in PostScript

PostScript Level 2 supports three different families of color spaces: CIE based color spaces, device color spaces, and special color spaces.

CIE based color spaces allow the specification of color in various CIE based models; the environment is programmable so that conversion of one model to another is possible. It should be noted that non-linear models can be supported. The color values are converted to an internal CIE XYZ system which can be mapped to a device color space by using the CIE based color rendering dictionary (CRD).

Device color spaces used in PostScript are depending on the output device technology. Models used for device color space are RGB for display type of devices, CMYK for color printing devices, and gray based color models for Black and White devices. PostScript also subsumes hue-saturation-brightness (HSB) models under device colors although devices are not addressed that way; HSB values are immediately mapped to RGB.

Special color spaces include color separation for process color printing, indexed color (which is widely used in Computer Graphics – initially only direct color was supported in PostScript), and pattern. Pattern is regarded as a distinct color space in its own right.

The illustrations 5.1 and 5.2 on pages 102 and 103 describe the different paths available for color specification and color rendering as specified in the "Red Book" [AD90].

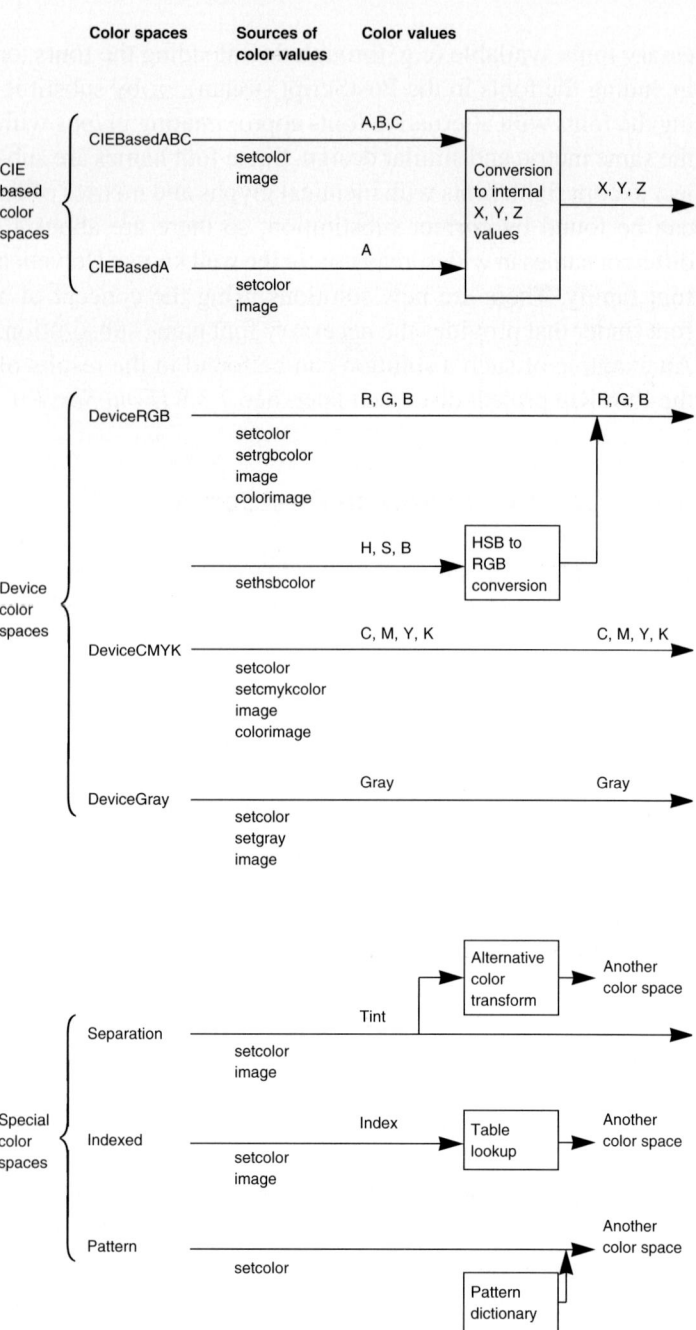

Ill. 5.1:
Color
specification in
PostScript Level 2

Incoming color values are converted to the color space of
the output device. After the conversion to device color space

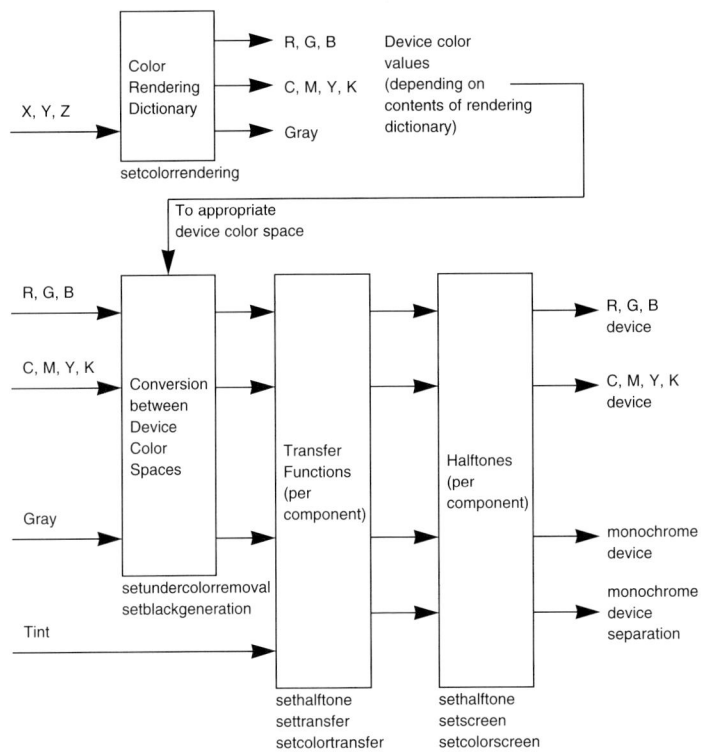

Ill. 5.2:
Color rendering
in PostScript
Level 2

transfer functions are used to allow for correction of non-linearities or polarity. If for a device the number of tonal values is restricted (e.g. to the two values ink and no-ink), at this stage halftoning techniques are applied to provide the necessary range of tonal values. Dominating technology for halftoning is amplitude modulated (AM) screening.

The area of screening is a critical one especially if it comes to separations for four color printing. As we have discussed earlier, screening angles and frequencies are especially critical in order to achieve a good quality result. Here PostScript Level 2 makes provisions for what is called accurate screening meaning a good approximation of the screening angles at the expense of processing time using the so-called super cell technology. Other methods of screening, e.g. frequency modulated (FM) screening, are not prohibited, but also not especially supported, however there are PostScript RIPs that also provide for FM screens under various brand names.

103

5.1.6 Device Independence

PostScript was clearly designed to provide device independence. Nevertheless it allows features that are strictly device or process dependent. Screening is one such example for device dependent facilities. Besides the fact that with Level 1 screening was used to realize device independent patterns, it really is a feature to simulate continuous tone output on bi-level output devices.

Many applications, especially the ones in the prepress context, often produce device dependent PostScript; looking at the printer selection menus gives a good impression of whether device independent code is really generated by that application. Many different PostScript devices typically show the utilization of certain features only available on certain PostScript devices. A large number of applications generate until today device dependent code using specific features which are not part of the PostScript language; these features are described in the device manuals.

Level 2 has provided some unification in respect to the necessary device specifics like paper formats or multiple trays.

5.1.7 PostScript Language Excursion

PostScript is a page description language (PDL). It has various aspects which we will discuss separately: PostScript as a language, as page description, and its basic graphics concepts.

5.1.7.1 PostScript as a language

PostScript can be characterized as a stack oriented language; it uses postfix notation of operators, i.e. operands precede the operators. The language is processed sequentially on a token by token basis. A PostScript file is therefore always a program normally encoded in ASCII. All data accessible to the program (including procedures being part of the program itself), are generated in form of objects. Every object has a type, attributes and a value.

We distinguish simple and composite object types. Simple object types for example are `boolean`, `integer`, `mark`, `name`, `null`, `operator`, or `real`. Composite object types are `array`, `dictionary`, `file`, `string`, and `procedure`; in PostScript Level 2 objects of type `gstate` and `packedarray` were added; in Display PostScript the types `condition` and `lock` were added. The following table shows some of the types with some examples.

Simple objects	Examples
boolean	`true false`
integer	`37`
real	`3.1415`
name	`/FontDirectoy`
operator	`add fill`

Examples of simple objects in PostScript

Composite objects	Examples
array	`[ 1 2 3 4 ]`
dictionary	`« /A 65 /B 66 »`
string	`(char string)`
procedure	`{ add dup mul }`
file	`♯stdin`

Examples of composite objects in PostScript

An object can be either literal or executable; this is an attribute. If the interpreter encounters a literal object, it is treated strictly as data and is pushed onto the operand stack; if it is executable it is pushed onto the execution stack and the interpreter executes it. For some types, e.g. integer, this distinction is meaningless; execution of an integer means also pushing it on the operand stack.

Access to objects may be restricted; so, for example, fonts are typically restricted to execute only. As is typical for a programming language PostScript provides programming constructs like `if` and `ifelse` in addition to an extensive procedure concept which heavily draws on PostScript dictionary functionality explained later in this chapter (see 5.1.7.4 *The PostScript Dictionaries*).

5.1.7.2 PostScript as a Page Description

PostScript is not only a programming language, it also provides mechanisms for rendering, for placing marks on a page. A PostScript file is a program describing the graphical content of a page in a device independent way.

5.1.7.3 Basic Graphics Concepts of PostScript

PostScript differs from other graphics formats in the basic concepts. Whereas most graphics formats employ so-called primitives, whose rendering aspects are determined by attributes bound to these primitives, PostScript uses a different approach.

The most elementary graphics concept in PostScript is the path. A path is a sequence of path elements, like straight line segments, second order non-rational Bézier curve pieces, and circular arcs; they have no immediate rendering, but rather define the underlying geometry. A path may contain disjoint subpaths, and a path may be self-overlapping.

This path can then be used by a series of operators for different purposes:

- rendering the path as a line using the `stroke` operator;
- filling an area described by the path with a certain color or pattern using the `fill` or `eofill` operators;
- using the path to define the new clipping region to which all output generated shall be clipped using the `clip` or `eoclip` operators.

The rendering operators each use their appropriate attribute values to define the exact result of the rendering.

One attribute common to all above mentioned rendering operators is color. Color includes normal color as well as patterns which are conceptually a special kind of color, as was pointed out earlier.

The `fill` and `eofill` operators only use the attribute color for rendering.

The `stroke` operator e.g. has the following rendering attributes associated with it:

- stroke width

- line cap (what exact form the end of a line should take, e.g. round or square)
- line join (how exactly the joining of two line segments in a point should look like, e.g. like drawn with a circular pen, with a square pen, or beveled)
- dash pattern (whether a line shall be drawn as a straight line, or dashed, and if dashed, the length of the dash line and space segments)

The difference between `fill` and `eofill` is the definition of what is interior and what is exterior.

The `eofill` operator uses the common graphics definition (even-odd fill rule) which makes a point inside the area if an arbitrary ray drawn from that point to infinity crosses the boundary an odd number of times.

The `fill` operator alternatively uses the so-called non-zero winding number rule. Again a ray is drawn from the point under consideration to infinity, and a counter initially set to zero is increased every time a boundary crosses the ray from right to left, and decreased, every time a boundary crosses from left to right. If the resulting counter is non-zero, the point is inside.

The same rules apply also for the `clip` and `eoclip` operators. Clipping in PostScript works cumulative, i.e. the pre-

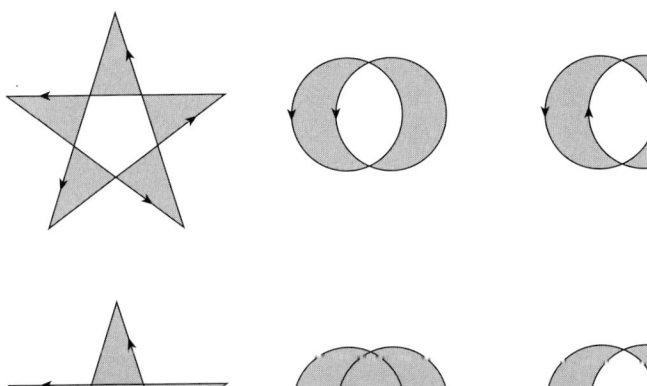

Ill. 5.3:
Even-odd rule
for filling

Ill. 5.4:
Non-zero winding
number rule for
filling

vious clip area is combined with the new clip area specified in a clip or eoclip operator invocation. Initialization of the clip area is to the complete page.

Other objects on the page can be text, generated from procedures for each character glyph, or continuous tone images or image masks. A whole range of attributes controls the rendering of objects in detail.

5.1.7.4 The PostScript Dictionaries

An important concept that has found usage also outside of PostScript is the concept of dictionaries. A dictionary is a named or unnamed collection of pairs of name and value, where a value can be arbitrarily complex; for example it can again be a dictionary, or, e.g. a procedure.

Dictionaries are very powerful means of specifying complex pieces of information. So e.g. ICC color profile functionality can be implemented in a representation of dictionaries.

5.1.7.5 The PostScript Execution Machine

As mentioned at the beginning PostScript is a stack-oriented language. The execution environment of PostScript provides several stacks for different purposes:

- The operand stack takes up data tokens (literal objects) from the input stream prior to their consumption by operators; e.g. if an input stream contains the sequence '3 5 add' the operands 3 and 5 are pushed onto the operand stack and later on consumed by the execution of the add operator which again itself places the result 8 onto the stack.
- The execution stack holds the nesting of the current execution context, i.e. the procedure nesting. This typically includes a job loop and certain error contexts.
- The dictionary stack holds the currently active nesting of dictionaries; the sequence from the top of stack down is the search sequence for named objects.

Besides the stacks the PostScript execution environment contains some more data structures related to the graphics processing. They are all collected in the graphics state. The graphics state contains all attributes and their current settings, the currently active clip path, and the current path to be used with stroke, fill, eofill, clip, or eoclip operators.

The tokens coming from the input stream are pushed onto the operand stack if their type is literal; if their type is executable they are put on the execution stack and executed.

The PostScript language has means of accessing the file system in the environment; during the initialization phase it reads from the environment files that define the startup state. After the initialization it typically starts the job loop and waits for jobs to rip. During idle times some RIPs perform time consuming calculations to prepare results that are likely to be used later. One such calculation is the scan conversion of the most often used fonts at common sizes; the results of this process is stored in cache memory. Caches of different sorts can also be part of a PostScript execution environment.

5.1.7.6 PostScript File Structure

PostScript files typically have a common structure: they start with a prolog which contains all commonly used procedures; often here is also a specific error handling procedure located.

After the prolog follows the description of the pages. The pages are ideally independent of each other. This however cannot be enforced by the PostScript language.

In order to make scanning e.g. for routing or scheduling purposes simpler certain comment conventions have been set up. The framework for such comment conventions are the Document Structuring Conventions (DSC) as given in the PostScript Language Manual [AD90] and elsewhere [AD92].

In some cases these comment conventions need to be followed exactly to allow proper processing of the file. Typical examples for these comment conventions are the rules for Encapsulated PostScript (EPS) [AD92c] files, the rules for Open Prepress Interface (OPI) [AL93], or the rules for Desktop Color Separations (DCS). The prologs in PostScript files are typically generated by applications, i.e. they are application specific. The most common thing almost any application driver does first in the prolog is to define all the lengthy operator names to single or two letter abbreviations. This significantly reduces the size of the PostScript code generated for the following pages.

5.2 Acrobat and Portable Document Format

In late 1990 Adobe announced a new technology called Acrobat [AD90b]. It is based on PostScript, and contains other information especially useful in display oriented environments. The technology and format behind Acrobat is called Portable Document Format (PDF) [AD93, BI96]. It is essentially a "flat" PostScript, which means that no procedures and control constructs are supported. This is probably mostly for efficiency reasons, i.e. to be able to do fast rendering to the display. The omission of procedures from PDF also causes a severe drawback compared to PostScript Level 2: in its first version it did not support device independent color, since it needs color rendering dictionaries with procedures to do the device color mapping; PDF Version 1.1 however provided a solution to this problem. Dictionaries are used in the PDF format to describe information that otherwise would need the concept of named variables. By this mechanism the original shortcomings regarding the device independent color spaces in PDF have been overcome.

Early versions of PDF did not support separations, i.e. it was very difficult to be used in high quality printing environments. Later versions however do overcome this deficiency. PDF also includes solutions to the font substitution problem based on the Adobe technology of multiple master fonts [AD92a]. It also utilizes the PostScript Level 2 filter and compression features to minimize the amount of data needed for representation.

Meanwhile PDF has become a significant component in the area of electronic documents. The original intent may be sacrificed to cover a larger area of application of this format. It is still not quite clear which direction PDF development will take in the future; there are too many interesting possibilities, and it is not clear which possibilities will be realized in what time frame. PDF and related products will remain an interesting area of development on which we should keep an eye; the more so as latest statements of Adobe announce a combination of PostScript and PDF called PostScript Level 3.

5.3 Tagged Image File Format (TIFF)

TIFF is a very common format for raster images once developed by Aldus [AL92]. The first version of the TIFF specification was published in 1986; revision 6.0 is the current version number. It has found wide acceptance as a de facto standard for image interchange in various application domains. Many applications allow to import and export image data in TIFF format. As a tagged format it provides a set of tags to supply attributes and image data values as tag-value pairs. TIFF is a binary format and supports the two major binary platform dependend variations distinguished by byte order and named according to the respective microprocessor manufacturers Intel and Motorola.

The TIFF document distinguishes two parts, baseline, and extensions. In the baseline part those features are described that all general-purpose readers should support. The extensions specify features that can be used by special or enhanced applications. For example the CMYK color model use is defined in the extensions, and in that sense, all prepress applications are special.

A TIFF file is defined as a sequence of 8-bit bytes. Its maximal length is 2^{32} bytes. It is structured and contains a header, and at least one image file directory, and image data. The header contains a signature that allows to distinguish whether the data are stored in Intel or Motorola binary format, a version number, and an address to the first image file directory. More than one image can be represented in one TIFF file.

5.3.1 Color Concepts in TIFF

An important criterion for image interchange is the capability with respect to color. TIFF supports bi-level, grayscale, palette (or indexed) color, RGB in the baseline TIFF, CMYK and various CIE based color images in the TIFF extensions.

111

Since also colorimetric data are available in the extensions, the color information exchange capabilities within TIFF are well developed. Whether a specific application can generate the requested colors depends on whether the implementers have taken the effort to implement it. This is to say that the capabilities are specified in the document, this does not necessarily mean that all tools implement these capabilities and so make these features accessible.

5.3.2 Image Organization

A common drawback for image interchange between different environments is the image organization. Scanning order is critical for rearranging data. If the ordering is different from the one necessary for the application, this can be quite time consuming, and is reducing productivity. This is especially true if the image is very large and/or has a very high resolution, a case that is almost always true when repro images are involved. A typical A4 full color repro image is in the order of 30-80 MB.

TIFF allows for different ordering or interleaving schemes of the image data. It also allows for segmenting images into small rectangular subregions – so-called tiles – in order to improve accessibility in different orders.

5.3.3 Compression and Encoding Methods

With high quality images the amount of data is always a problem. Therefore the utilization of compression mechanisms is of great importance. TIFF supports a variety of methods for compression and data reduction.

Besides normal runlength encoding (called PackBits compression), Lempel-Ziv-Welch (LZW) [WE84] compression, facsimile group 3 [CC85] and group 4 [CC85a], as well as JPEG [IS92] baseline compression (using discrete cosine transformation) together with various encoding methods like Huffman encoding can be used to reduce the size of images.

5.4 CGM

The Computer Graphics Metafile (CGM) [IS92a] is a standardized means for exchanging picture information. It has been widely adopted by industry following its inclusion into the US Department of Defense's CALS initiative. Many applications today provide results in form of CGM files. The standard comprises four parts, one functional specification and three encoding parts.

5.4.1 CGM 1987

The first CGM standard was published in 1987. It contains basic functionality utilized in graphics systems excluding structured objects like segments or PHIGS structures. It's concepts are derived from graphics systems with the intention to allow pictures to be exchanged between applications and sites. In context of document applications the concepts are rather Spartanic. This is to say that document applications often require closer control over the appearance of an object (e.g. line ends, line joins) than was provided. This and the lack of structuring facilities lead to several amendments which were incorporated in a new edition of the CGM standard published in 1992.

5.4.2 CGM 1992

This new version of CGM brings many improvements, and it is well prepared to compete with page description languages with respect to their functionality. In some areas the new CGM is superior in capabilities even to PostScript. It allows control over all the graphics attributes also available in PostScript, and it provides facilities missing from PostScript, e.g. filling areas with color interpolated across the area. This feature is very common in DTP systems, when you generate a background for a page having dark color or gray at the bottom and light color at the top of the page.

An interesting aspect is the handling of fonts. CGM allows to specify a list of fonts to be used, and allows to select the use of a certain font by its index. This allows to provide CGM pictures with good quality fonts. However, it poses the problem to CGM interpreters that they should provide high quality fonts and font rendering.

5.4.3 CGM Encodings

As mentioned above CGM defines three different encodings, a binary encoding, a character encoding and a clear text encoding. Each of these encodings has a distinct purpose. The binary encoding is intended for maximum efficiency encoding and decoding a CGM file; it utilizes a binary (machine) representation that in many cases is identical to the binary data types of the computer utilized. The character encoding is intended for use for storage and in network environments where it is important to have compact data representation to avoid too much transfer/reading time. The clear text encoding is designed to allow human readability and manipulation with a simple text editor.

5.5 Image Interchange Facility (IPI-IIF)

IIF, the Image Interchange Facility [IS93] is part of a standard named Image Processing and Interchange (IPI) developed by ISO/IEC JTC1/SC24 Computer Graphics and Image Processing. It broadens the concept of images to also include time varying images; in fact it generalizes images at least conceptually to n dimensions, thus including not only time varying but also multi-channel data and even 3D volume data varying over time in the same model. IIF is the interchange part allowing ASN.1 encodings to be used, the encoding method dominant in the PTT world of interchange.

The whole IPI standard, including IIF is designed to be a generic standard, i.e. a valid standard across domains, not domain-specific. For different areas application profiles are expected to develop in order to specify which facilities are selected for the specific application domain. IIF is a powerful format for future use also in the document domain. Its development has just been completed, but we are still lacking products supplying IIF format files. There are some pilot implementations already available that even support multimedia image exchange. IIF is the basic image interchange format in several national and international projects, dealing with multimedia mail. This is an important step towards an open image communication architecture.

5.6 Prepress Interchange Formats

Prepress interchange formats are extremely vendor specific. They typically contain at least three kinds of data: continuous tone images, line work or line art (in Computer Graphics terms indexed color images) and bi-level images, often called masks. The resolution of continuous tone images is about 120 lines/cm, with line work and binary images the resolution is typically from 500-1200 pixels/cm.

5.6.1 IT8

IT8 is the American National Standards Committee for the graphics arts. IT8 has issued several standards for image interchange [AN88, AN88a, AN90]. These standards however have no practical significance. Except for a new development that is based on the TIFF format – the so-called TIFF/IT [AN93], the standards developed here are media dependent, in fact they are magnetic tape standards misusing labeled tape formats by putting data into labels, utilizing nested user header labels.

5.6.2 HandShake

HandShake [SC88] is the proprietary format of Scitex for prepress data exchange. It is used with Scitex prepress systems, and also is used sometimes as an exchange format to other systems. Continous tone color data are always modeled as CMYK. The format is relatively simple to generate and interpret, and therefore is well suited for exchange between systems. For continuous tone images it is restricted to 8 bit information per color component.

115

5.6.3 ChromaLink

ChromaLink [HE90] is another proprietary exchange format of the former Dr.-Ing. Rudolf Hell GmbH. ChromaLink, similar to HandShake, is a well suited exchange format. It is more complicated in its structures, but is read by several other vendors systems. For contone images it allows up to 12 bits per color component. It also specifies the integration of line work and continuous tone data with line work being supplied at a six times higher resolution than continuous tone.

5.7 PhotoCD

A standard medium today is the compact disk. Since Kodak introduced their PhotoCD product [KO92] high quality color images have seen a significant increase, because PhotoCD has achieved a reliable scanning process, one of the more critical steps in producing quality at a relatively low cost. PhotoCD includes an image file format that is proprietary; important parts in the processing of PhotoCD images are hidden. While the intent of Kodak is obvious, it is a step backward with regard to open systems. The toolkits to access PhotoCD images are available, so one can implement readers of PhotoCD quite easily; but if your application is modifying the images, you can go out to other formats only, except you are willing to pay license fees to Kodak for a library allowing to also write PhotoCD.

Another disadvantage is the discrete resolutions you can get on a PhotoCD; for some applications this will mean you will need higher resolution than your application requires, and will have to do image processing in order to get what you need. Still PhotoCD is a means of acquiring high quality color images at a rather low price. To achieve similar (and probably better) quality the only alternative would be to go to a repro scanning process, and this tends to be more expensive.

5.8 Document Related Standards

We see a proliferation of document related standards these days. Some of them are still in a rather early stage of processing. However some are gaining momentum already now. We will discuss here only those that already have a certain impact on practical work.

5.8.1 Standard Generalized Markup Language (SGML)

The Standard Generalized Markup Language (SGML) [IS86] is a language to specify a markup syntax. With SGML you can define the legal logical structure of a document in a so-called document type definition (DTD), and with appropriate editors you can ensure that only correct documents according to the given DTD are produced. SMGL has gained a significant momentum by the American Department of Defense through their CALS initiative, and it is currently being tried out in several contexts, including scientific papers for journals. Documents marked up with SGML are not final form documents, which is different from all the other formats we have looked at, but it contains all the logical information necessary. Together with a mapping of tags to a formatting system final form can be produced, and this final form will again be most likely in one of the formats discussed.

This mapping is supposed to be achieved through the - Document Style Semantics and Specification Language (DSSSL) [IS96].

A specific Document Type Definition in SGML is HTML, the Hyper Text Markup Language [SA96]. It is widely used in the World Wide Web (WWW) on Internet. Although neither SGML nor the DTD HTML are formats directly input to a ripping process, they have gained wide acceptance, and together with a dynamic formatting process they are able to generate input for ripping. This development also leads to a widening of ripping concepts to multimedia presentation.

5.8.2 Standard Page Description Language (SPDL)

After much discussion over many years the Standard Page Description Language (SPDL) [IS93a] is available, and it is effectively the semantics of PostScript Level 2 with different syntax. The question whether this will become a successful standard is very much depending on whether large organizations or governments will be backing it. We can rightfully doubt that it will have a successful future. It should however not be too difficult to implement it on whatever PostScript interpreters are available. It can definitely be realized as a set of PostScript procedures, provided it can build on top of a PostScript Level 2 interpreter implementation.

6 Problem Areas

Believing advertisements, problem areas are non-existent. Real life experience shows that there are still a few problems left to be dealt with. We would like to mention here mainly two areas of concern: color, and fonts. We will discuss both of them briefly to get an idea about the nature of the problems observed.

6.1 Color

The proliferation of relatively cheap color output devices is one reason, why color tends to develop into a problem. Especially color printing devices form a non-trivial task for ensuring the right color to appear on paper. As we have seen earlier, color is not a physical phenomenon, but it is a complex perceptual problem that involves many physiological and psychological processes. The modeling of these processes requires non-linear relationships, and it may even involve non-reversible mappings under some conditions.

6.1.1 Color Space

A few years ago it was sufficient to support a device dependent RGB color space. Sometimes we found some physiology oriented color spaces building on hue-saturation-brightness (HSB) or similar models. Meanwhile the application require ments go far beyond this simple model. A CIE-based color model is becoming a necessity, and this is even true for RGB-models which carry their CIE primary values, the CIE color coordinates of the three phosphors used to display.

6.1.2 Color Calibration

As long as we talk about displays, things are relatively simple. For displays we have means for converting CIE coordinates to RGB display values (as long as they fall within the gamut of the display) by simple matrix multiplication (and gamma correction). As soon as it comes to printing devices these simple mechanisms fail. Mapping a specific color to a printing device is a highly non-linear process. It is not sufficient to measure a set of points regularly distributed and then use linear interpolation to generate values in between these measured points, except for the case that the grid of measured points is extremely dense. Besides non-linearities the other critical issue is that there are no models describing the printing process to a degree that would allow building a program that would reasonably resemble the practical printing process, or allow predictions about the behavior under given conditions.

Calibrating a printing process with four process colors (CMYK) as they are used commonly in printing presses is a one-to-many mapping; as discussed earlier, the amount of Black generated and the amount of Cyan, Magenta and Yellow removed can be vastly different for the same input color. This makes the process of Black generation and under color removal in most cases an irreversible process since the parameters used in generating CMYK are generally not part of the information transmitted. Dyes used in printing clearly modify the final result. A digital image processed under the same conditions and printed in Europe and the USA will give different results, because printing inks in Europe use different dyes from printing inks in the USA.

6.1.3 Color Management

In order to cope with the problems just mentioned, color management tools have started to appear in the market. Color management is a means of providing uniform color across devices. The tools normally work well with displays – calibration is not a real problem there except for colors that lie outside of the gamut of the device – but have difficulties with printing devices or printing processes. Normally they produce acceptable results with a small number of devices (usually with devices of the same brand as the color management system), for which device color measurements are available, and provided the devices are reasonably stable over time.

6.1.4 Transparency

Another problem that typically arises when designing some piece of graphic arts with typical design software is related to layering different parts; the relation between two overlapping parts cannot be modeled according to the PostScript imaging model. This model allows only to set color values not allowing any combinatorial effect with what is already on the page. It is based on the concept of a page memory which can only be written. Sometimes one wishes to combine an object with its background thus giving the effect of a transparent overlay. This effect is generally not possible with today's page description languages. This has the consequence that one has to produce such a page on a repro system and import it later as a pure raster image. This is not only inconvenient, but it also restricts the output quality by the resolution of the raster image produced.

Alternative imaging models are necessary, but are currently not supported. Transparent overlays or – more generally speaking – combinatorial imaging models that also allow e.g. complementary setting of colors can easily be built into raster image processing software, and recent research results show that this is worth while doing.

6.1.5 Color Interpolation

A very common operation in desktop publishing systems today is providing a background that is interpolated between colors across the whole page. This is normally represented as a step function using varying discrete color values for filling areas across the page. This is not very annoying with low to medium quality output devices; they often cannot do better anyway. But with high quality output, e.g. on laser image setters for high quality press printing, this is significantly reducing possible output quality.

This problem has been covered by the latest developments in CGM, but it is still unsolved in relation to PostScript. Here also we expect extensions to be proposed that would properly solve this problem, and besides this would also reduce the amount of data necessary to describe the interpolated background. CGM clearly shows the direction to take, allowing different typical geometrical forms and ramp functions to be used for the interpolation.

6.2 Fonts

One of the quality criteria for a document output is the proper availability and use of fonts. Ten or fifteen years ago one could only select from a small set of sizes for a few fonts. With the advent of PostScript this situation changed dramatically. Nevertheless the font situation today is far from being to the full satisfaction of users. Copyright problems with font names and incompatibility between fonts is still a source of frustration when exchanging documents electronically. With the proliferation of Microsoft Windows the problems have grown worse, because yet another font format has come into play.

6.2.1 Font Formats

For many years PostScript Type 1 [AD90a] was the only widely available format for font resources. The drawback of Type 1 was at first the encryption for the fonts. With the opening of the description of a new format, the TrueType format and how to integrate it into a PostScript stream [MI90, AD93a], the politics behind the encryption of fonts was strongly challenged, and the result was the publication of the Type 1 format description.

Besides these two some more font formats are available, but these two are probably the most important. Type 1 definitely has the largest installed base and by far the largest set to select from, but TrueType is becoming more and more popular with the wide distribution of Microsoft Windows.

Typical software using TrueType fonts requiring output in PostScript format needs to generate the font information in the output stream. This however is often done by utilizing the Windows capability to scale and raster fonts, i.e. the fonts generated on the PostScript output are raster fonts! In order to have at least medium quality results in print the output has to be generated towards a fixed output resolution. On devices with higher resolution this will generate low resolution results for type. Quality font technology today is based on outline fonts using sophisticated hinting techniques to produce good quality raster representations; with already rasterized fonts this is not possible.

After opening the descriptions of different font formats one would assume that it would be possible to generate an alternate font description from one source font description. This is however not as simple as one expects. Besides the technical problems that are not easily solved, font description can be programs that are protected under the copyright law, which could make such a conversion illegal even if it were possible.

6.2.2 Font Exchange

Due to copyright it is problematic to supply the required fonts with the output file. Technically speaking fonts in the output file would enable the receiver to use this font also in other context than the document it came with. Something that is technically possibly may still be illegal. In order to inhibit this possibility often only font names are supported in the document. This leaves the receiver with the problem of allocating the necessary fonts to present the document properly. Just recently and under certain restrictions some fonts suppliers have agreed that it is legal to supply a font within an Encapsulated PostScript (EPS) file, but that it is illegal to use this font for anything else than generating the output.

For typesetter studios a strictly legal position results in a significant investment in fonts in order not to have too many surprises with documents requiring fonts not available. In many instances if a font for a document is not available the results are at best not good looking, if it comes worse, the document may be completely cluttered and unreadable. But even in case that a studio has legal copies of all the fonts required there is a chance that a font with a certain name may have variants that are not easily recognized, and that these variations may cause different typesetting results in print.

An alternative solution at the possible expense of quality to the font availability problem could be an intelligent font substitution.

6.2.3 Font Substitution

Font substitution used to be a non-issue because reasonable font substitution was not available. A typical fallback position in PostScript was using Courier for fonts not available if not completely skipping the job.

New technologies including the Adobe Multiple Master [AD92a, AD94a] and URW TigerType allow for finding better font substitution alternatives. If a font is not directly available a font is selected that has the same metrics and similar design characteristics, e.g. serif/non-serif, regular/italic style, consistent cap height/body height ratios, same weight, or same expansion.

With this technology it is possible to select a font for the requested one that gives an appearance of the document close to the one intended. This is still not the same as having the proper font available, but this is more likely to be accepted than using a completely different font.

A difficult problem that normally requires reformatting a document is the change to another font family. But even this difficulty can be overcome by font families specifically designed to use the same metrics across multiple families. Thus digital type can be varied not only in height, weight, and expansion, but also in design.

A practical issue however should not be skipped here, which is the naming problem of fonts. For the well known non-serif font Helvetica (which is a registered trademark name) there are more than 40 different alternative names in use. Similar statements hold for other well-known font names. Often the problem is not so much to obtain the right glyphs and metrics for a font but rather to find out which font was really meant to be used. To know that Holsatia, Swiss, Helvetstar, Helvetia, Triumvirate, Europa Grotesk, Newton, Spectra and NimbusSans all mean the same font requires knowledge about fonts and font names that is not present everywhere.

This implies that a font name database is a necessary requirement for proper font substitution. Optimally this database should not only contain the name equivalencies, but also the design characteristics of each font so that an intelligent substitution can be made if there is no exactly matching font available. See also 7.3.8 *(Font Server)*.

6.3 Resource Requirements

Interchange of electronic documents in electronic form is common place today. Whether one can present an electronic document properly is a question of availability of resources required for presentation. We can distinguish between two kinds of requirement, one kind needing special hardware properties (e.g. full color output required), the other kind needing special software including fonts or special format converters.

For PostScript output it is a well established practice to provide resource information at the beginning of a file so that e.g. spooling systems can decide to which printer to route a specific file. This information may also be used to pre-load certain fonts not available by default. Other formats like CGM have a standard practice to provide a record that specifies which functions are to be expected in the file; short hands for very common functionality sets are also available.

Specifying resource requirements should develop as standard practice for all kinds of formats. Versions of certain formats should be indicated as well as special needs on software and hardware.

7 The Integration Problem

In the following sections we will discuss possible solutions to the need for integrating various formats into one document, maybe even on one single page.

7.1 Integration of different Formats

Having discussed all the different formats the question in the end is: how can I integrate all these formats? Except for the discussion with PostScript and Encapsulated PostScript we have not dealt with this question up to now. Obviously there is the one alternative that was pointed out during the EPS discussion: only generate EPS files for objects to integrate, and for the main body of the document generate PostScript. This will solve the problem. Another alternative would be to provide tools that allow the switching of context during ripping and thus allow direct integration of the various formats on the page. Both alternatives will be briefly discussed here.

7.2 Conversion to a single Format

Looking at current practice in the field it is unrealistic that every application generates one single format. What really happens is that a large set of converters between formats exist. To simplify our discussion let us assume the format generated during conversion be EPS. After converting all foreign formats to EPS the different files are merged together with the main body of the document. This solution is shown in the following figure.

This approach has some significant drawbacks. At first every single entity to be integrated exists three times, as original in whichever format it was, as EPS file after conversion, and in the integrated file that is finally going to be printed. This means a lot of disk space is being wasted. The second problem is that not all information in the original file format may be mapped to EPS. This is even more severe because we may end up with a semantic degradation that is not acceptable.

Therefore another solution is necessary to avoid the above mentioned problems. This solution requires integration of different formats without conversion.

Ill. 7.1:
Conversion to a
single format

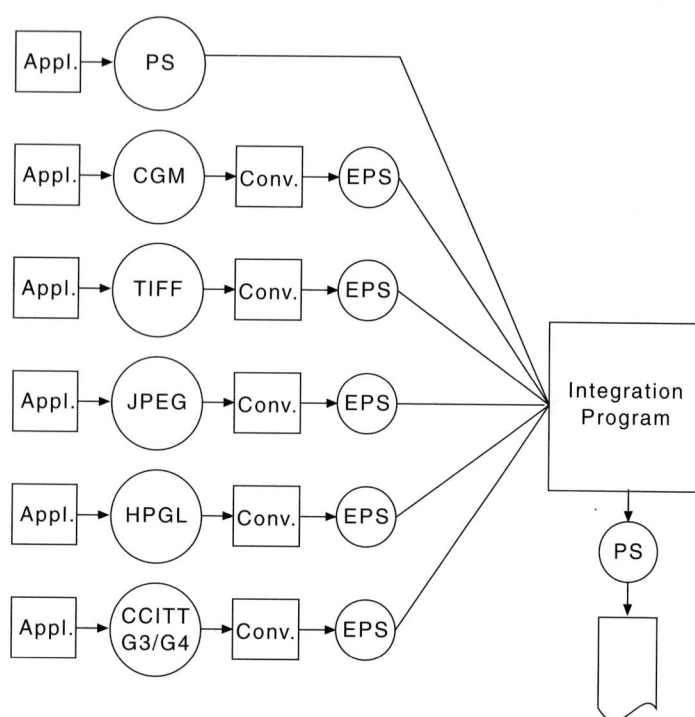

7.3 Integration without Conversion

Currently we already find a number of printers that can emulate different formats. The drawback of those devices is that they can only handle one format at a time and for a page. In order to overcome this limitation an architecture was developed for an Integration Raster Image Processor (Integration RIP).

This development was done in the framework of the ESPRIT II project 5167 EuroRip [WI91]. A brief description of this project and its results follows.

The main goal of the EuroRip project was the seamless integration of content represented in several page description languages, graphics languages, raster image formats, and font formats on all standard platforms. The conceptual environment is shown in the figure 7.2 *(Integration without conversion)*.

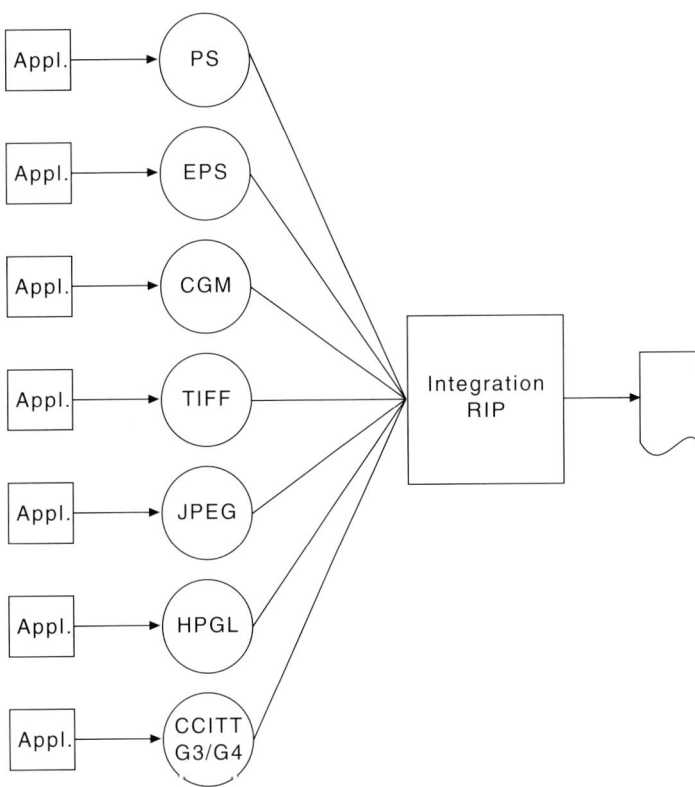

Ill. 7.2:
Integration
without
conversion

lems related to the use of fonts. Therefore a variety of standard formats were supported: PostScript Type 1, PostScript
Composite, TrueType, and Nimbus (the native format of the
project partner URW). In the Font Finder database, many font
names known today are stored and mapped either to the original installed font, to a font equivalent to the original one, or to
a similar font selected from the 500 TigerType typefaces. A
font mapping example is given in the figure below.

Ill. 7.3:
Font Finder
mapping example

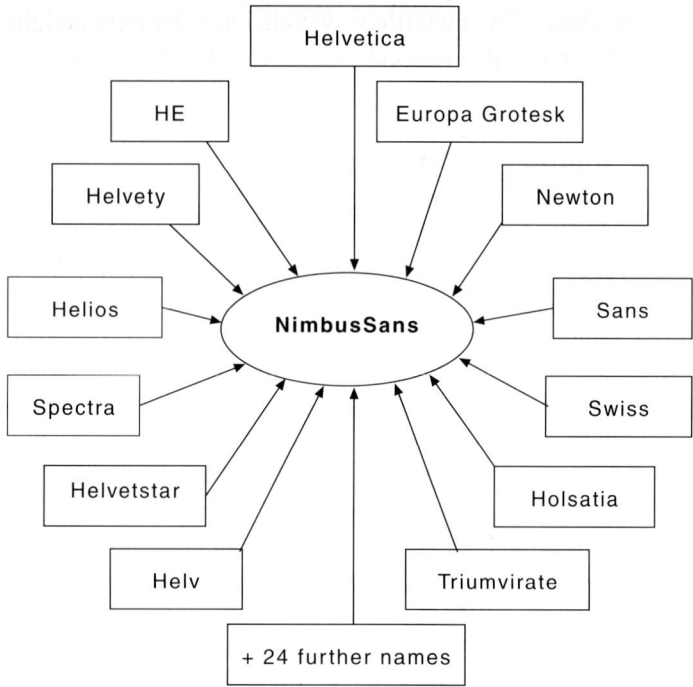

7.3.9 TigerType

TigerType was a set of 500 fonts for all common needs. It
covered ten typeface families each consisting of 50 fonts in
different weights, 25 in regular style and 25 in italic style.
TigerType was used as the font base of the Font Finder database. If a certain font was required by the interpreter kernel
but not installed in the font server, the Font Finder looked for
a similar font with the same x-height, stem width, and total
width within the 500 TigerTypes.

8 Information Interchange for the Production

Interchanging information for document production is a critical area. Mechanisms for interchange of information must fulfil certain criteria in order to be useful:

- the mechanism must guarantee the integrity of the information;
- it must guarantee the consistency of the information;
- it must guarantee the completeness of the interchanged information;
- in shall ensure that the information is self-contained;
- it must ensure that the information has no side effects;
- it must ensure that there is no (or at least no unacceptable) loss of information.

In order to allow better insight into the interchange problems we will look at a small set of typical scenarios from practical environments.

8.1 Media Use for Information Interchange

In the past the interchange medium was mostly film or paper. With the success of electronic prepress systems, and with high quality requirements, the use of electronic media for transporting information became dominant. Here cartridges of different types became important. The most critical specification values of such cartridges were the data capacity and the speed to transfer data to or from the cartridge. Both disk and tape cartridges are in use, where disk cartridges are more often used for information exchange, while tape cartridges are more frequently used for backup.

With the advent of write-once compact disks this medium became also used for information exchange in the prepress word. Magneto-optical disks also became widely used for that purpose.

Besides media where something has to be physically transported, new media technologies are becoming available through networks. Network systems are widely used worldwide today. The rapid growth of the Internet and of World Wide Web are indicators for the popularity of electronic networks. Sending artwork information across networks however requires high bandwidth communication, and this is still not available at reasonable cost. The cost/performance ratio will be a key factor for success of networks in this area in the future.

8.2 Interchange of Editable Objects

For the interchange of editable objects today still in most cases application dependent formats are being used. The previously mentioned problem of having identical versions of the application at both the sender's and the receiver's site is crucial here. However there are signs showing that other methods are also possible.

For image interchange the TIFF format has been in use for quite a while. This format is application independent, and is an editable pixel format.

But also for graphics there are approaches to allow application independent interchange. Examples are the Computer Graphics Metafile, the Adobe Illustrator subset of PostScript [AD92d], and several approaches based on distilling technologies for PostScript files (Adobe Acrobat [AD93, AD94]).

PostScript editing technologies coming into existence are useful with restricted types of PostScript sources; they are critical with regard to arbitrary (perhaps hand coded) PostScript files.

8.3   Advertisement Transmission from Prepress to Newspapers

The times when newspaper ads were transmitted as film are fading. Today it is necessary to find solutions to transmit ads to a newspaper electronically. The transmission of advertisements underlies the above mentioned criteria. The relations between the participating parties - agency, repro house, and newspaper – are manifold. One concrete situation using the Advertisement Transmission System (ATS) is shown in diagram 8.1.

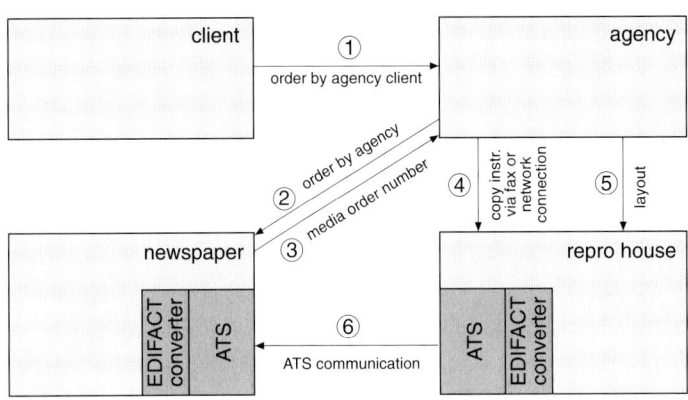

Ill. 8.1:
Relation between
agency, repro
house, and
newspaper

The format chosen to transmit the ad must ensure some of those criteria, some must be ensured by the transmission mechanism. A good example for such a transmission mechanism is given in the *IFRA Special Report 6.14.4 -Recommendations for basic requirements of Artwork Transmission Systems* [IF95]. The requirements listed there are
- platform independence;
- integration of administrative data;
- use of application-independent interchange format;

These criteria are fulfilled in the following manner: The format used for embedding the artwork is either Encapsulated PostScript (EPS) or TIFF. These two formats guarantee application-independence. The use of these formats also provides an important contribution to platform independence.

Application dependent formats (the application internal formats) can be useful only in very restricted environments,

135

e.g. one must guarantee that both sender and receiver use exactly the same version of an application, and the font environment must be the same. Otherwise no consistent transmission between different environments can take place.

The transmission itself is based on Internet Protocols, on the File Transfer Protocol (FTP) in particular. This makes the transmission independent of by what means the connection is made; it also makes it independent of the platform used, since it is implemented on practically all platforms imaginable. FTP is also used to ensure the integrity of the information transferred.

Ill. 8.2:
Layered
architecture of
ATS

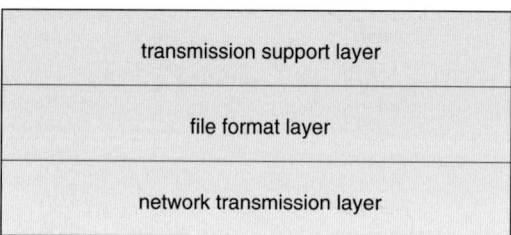

In order to fulfil the other criteria mentioned above some further restrictions have to be placed. So it is not possible to maintain file references (e.g. to high resolution image data) in the EPS file; these references have to be resolved and the data included in order to make the file self-contained.

The allowed compression mechanisms in the files are restricted (runlength, LZ77 [ZI77]) to avoid loss of information.

In order to guarantee the completeness of the information, all fonts used in the ad must be supplied within the EPS file. The font problem (already dealt under Problem Areas) cannot be solved satisfactorily in another way because even the same font names do not always guarantee identical fonts, and in an advertisement fonts are also used to convey part of the corporate identity that needs to be preserved. This also contributes to the consistency of the transmitted information.

Another component to consistency is achieved via the added administrational data. These give e.g. the reserved space for the ad in the newspaper, and this can be checked against the size of the contained artwork data; this checking allows to recognize inconsistent size information before sending.

The requirement to avoid side effects is partly fulfilled by the EPS specification; e.g. it forbids PostScript operators that change global states. Also the embedding rules for EPS files ensure that there are no side effects produced from the file that have effect outside of the file.

ATS allows to transmit finished objects that are ready to print. These objects are typically not editable; for the interchange of editable objects other methods have to be chosen.

An ATS network provides for three types of partners:
- ATS sender (agency or repro house),
- ATS server (router service), and
- ATS receiver (newspaper)

Diagram 8.3 shows possible network connection configurations.

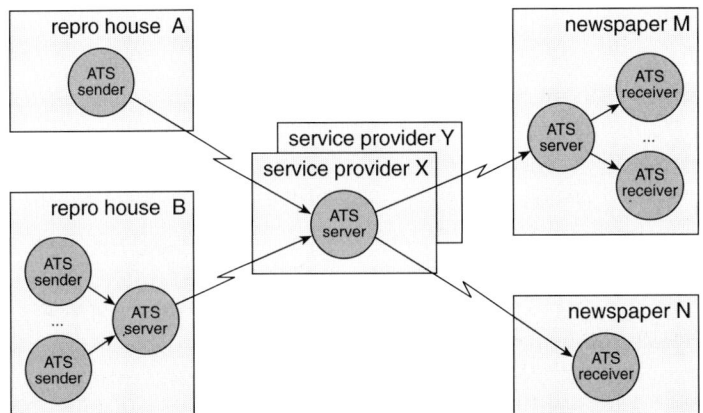

Ill. 8.3:
Possible ATS network configurations

This allows an advertisement to either be sent directly from the repro house to the newspaper; if it is sent to multiple recipient newspapers with common paths, it can be routed through servers that scan the address information and ship the respective informations on to other servers or one of the receiving newspaper.

A proof of concept implementation of ATS has been demonstrated on Macintosh, Windows´95 and Windows NT platforms by the Fraunhofer Institute for Computer Graphics.

8.4 The CIP3 Print Production Format (PPF)

One objective of new developments for the production of documents in print is the reduction of make-ready times. With digital presses coming into play time reduction becomes even more demanding. If we are able to print sheets at high frequency with low volume, going down to one copy, with content change from sheet to sheet, not only the time to print counts; the rest of the system has to be integrated and streamlined too. Production direct to delivery squeezes setup times in the press and postpress area to almost zero, This can only be achieved by a fully digital link of all production steps. The goal of all these efforts is to boost productivity.

8.4.1 What is CIP3 PPF ?

CIP3 stands for *International Cooperation for Integration of Prepress, Press, and Postpress*. PPF stands for Print Production Format. All that means is CIM for print products. The first step was taking information available at prepress to press and postpress stages. The underlying format PPF uses PostScript as the encoding method.

8.4.2 How to Boost Productivity?

One possible way of boosting productivity is to reduce production steps that may no longer be necessary. Another important action is the elimination of manual interaction steps and substitute them for automated process steps A further boost to productivity can be to speed-up individual production steps.

Altogether we need to automate production, build a computer-integrated production for print products.

The following production areas serve as examples of the topics that need consideration; they are by no means an exhaustive list, but give a good impression of work to be done.

8.4.3 Order Processing

In times when a print shop could survive in the market with a few (e.g. five) jobs per day, orders could be delivered with conventional means; for example orders came in by a courier delivering data on a cartridge with some paper describing the details of the order. In times of short run color print, with short run meaning something between one an a few hundred, it may be necessary to have a throughput of 500 jobs per day to survive. With 500, or even with 100 jobs per day a conventional delivery of orders can no longer be achieved. In the future orders will have to be sent via network. This means that all information, artwork and description of the order have to be sent electronically.

8.4.4 Prepress

The complete artwork production is probably the area where most automation steps have already been implemented. Networks are already common in this area. That means the prerequisites are already available for sending artwork information accompanied by order information via network. Smooth workflow however is still a rare and exceptional case.

8.4.5 Press

One issue of the future in printing is individualization of products, be it single products manufactured to the individual needs of the recipient, or be it sets of products manufactured to the need of a certain group of recipients.

This requires printing to be integrated with the following finishing steps.

8.4.6 Postpress

In line finishing of the components to the end product is an important step towards streamlined production of print products. The result of that step is a: print product ready for delivery.

8.4.7 Delivery

The automatic feeding of products to a courier service or some other mail service is a consequent step to finalize automated production. Distributed printing together with delivery arrangements are the base for wide area support with printed material.

8.4.8 CIP3 History

A brief history may help to understand the developments of CIP3. Back in December 1993 the idea of a smooth integration of production was discussed by people from the R&D of Heidelberger Druckmaschinen AG and from the Fraunhofer Institute for Computer Graphics Department "Document Imaging". A study was performed by the Fraunhofer group which lead in September 1994 to a first internal draft specification. In December 1994 a first functional prototype was available that could demonstrate some of the potential of the proposal. In February 1995 the foundation of CIP3 group was organized by Heidelberger Druckmaschinen and Fraunhofer.

In May 95 the version 1.0 of CIP3 PPF [CI95] was published and presented at DRUPA 95. In August 1995 the first CIP3 PPF file was created by a prepress product beta version. In August 1996 version 2.0 of CIP3 PPF [CI96] was published.

Throughout 1996 many field tests were deployed and experiences in practical applications were gathered.

The CIP3 consortium expanded from initially 15 to 26 member companies at the time of writing.

8.4.9 Members of CIP3 Group as of May 1997

Adobe (US)
Agfa (B, US)
Baldwin Technology Company (US)
Barco Graphics (B)
Creo (US)
R.R. Donnelley (US)
Ekotrading-Inkflow (SLO)

Eltromat Polygraph (D)
Ewert Ahrensburg Electronic (D)
Fujifilm (former Crosfield) (UK)
Goebel (D)
Harlequin (US, UK)
Heidelberger Druckmaschinen (D)
König & Bauer-Albert (D)
Kolbus (D)
Komori (JP)
Linotype-Hell (D)
MAN Roland (D)
Mitsubishi Heavy Industries (JP)
Müller Martini (CH)
Polar-Mohr (D)
Scitex (IS)
Screen (JP)
Ultimate Technographics (CDN)
Wohlenberg (D)
Xerox (US)

8.4.10 Content of CIP3 PPF (V2.0)

Print Production Format (PPF) files contain the following information:
- Administration data is contained in the PPF file to allow the link to job administration systems that may be
  installed in a shop.
- A low resolution image for all separations is part of the
  file; it typically has a spatial resolution of 2 pixels per
  millimeter, and a tonal resolution of 8 bits per pixel.
  These images can be used to calculate ink key presetting and ink consumption from the electronic file.
- Transfer curves describing the process can also be part
  of the files; they are needed to make correct ink calculations.
- Register mark positions are also stored in the PPF; they
  can be used for automatic registration of the press.
- Color and density measurement can be specified in PPF;
  they can be used in an automated quality control
  system allowing a feedback loop to the process with

141

requested color or density values defined in the PPF file.

- Cutting and folding information can also be transferred via PPF. After imposition this information is known, and it is made available through PPF. Of course cut and fold information in the PPF shows ideal measures, but they can serve as a good starting point for automated exact measurements.
- Besides these information components other information can be stored in private data.

To show the differences between state-of-the-art in print production and the application of CIP3 PPF the production of a printed sheet is reviewed.

8.4.11 State-of-the-art Example

After all creative steps have been done, imposition is made and the imposed job is sent to a file. This file is ripped and the film is made. With this film we can now copy the plate. The pressman takes this plate and reads it on a plate reader, thereby scanning the image on the plate. This information is now used for the calculations for ink key presetting. Then the plate is mounted, the press is preset, and the press run started. The ink key setting is controlled and adapted as necessary. Registration of the press is controlled visually, and manual corrections are performed. The result is a stack of printed sheets.

8.4.12 CIP3 Example

Let us have a look at how the same process is executed using CIP3 PPF.

It is important to note here that the application of CIP3 PPF requires a fully digital imposition of the sheets. The information generated during imposition (and possibly during the RIP process) are used to calculate the ink key presetting for the press. This can be done in parallel to the generation of the plates, since the plates themselves are not required for the calculation. In the traditional process plates have to be ready before continuing. Besides digital information can be used for register setting, as well as also for color or density measurement and control. The result is also a stack of printed sheets.

Comparing the two processes it is obvious that the use of CIP3 PPF – when supported by products – can be much faster, cheaper, and better quality controlled.

8.4.13 Current Use at Different Stages

Taking the three major stages prepress, press, and postpress, we can envisage the following uses of CIP3 PPF.

Prepress uses CIP3 PPF specification in imposition software, and perhaps, when ripping to generate the PPF files.

Press use is the first PPF consuming step; it can utilize the information in PPF for ink key presetting, as well as for register control, and color quality control

Current postpress use is possible for cutting and folding.

8.4.14 CIP3 PPF Extensions in Work

Meanwhile development goes on, and several PPF extensions are currently in work.

Up to version 2.0 of PPF only a single sheet was allowed on one file; from version 2.1 on multiple sheets can be stored on one PPF file.

Also part of development are the definition of the finished product (V3.0), including further postprocessing steps to describe what for an end customer is the product, e.g. a brochure, or a book.

The finishing steps included in version 3 will be among others collecting, binding, trimming, and insertions.

8.4.15 Productivity Benefits of CIP3

Looking at the overall benefits in using CIP3 information the following observations can be made:

There is no more duplication of data acquisition, as has been the norm in conventional processes. A result of that are shorter production cycles, better quality control, and lower cost. With CIP3 PPF a streamlined production is possible. CIP3 PPF is an important information container used in production flow.

8.4.16 Experiences in Practice

The first field test installations have given positive feedback from practice. Current field test installations make use of CIP3 PPF for ink key presetting; the quality of preset values is so high that preset values can be used directly for production run without further adjustments of the keys.

Savings of press time are significant; the reported savings are up to 80 minutes press time a day. With the high investment for a press this is a signifiacnt amount of savings.

8.4.17 CIP3 Environmental Benefits

Besides productivity also the environment benefits from the CIP3 approach. The fully digital production required also forms the basis for other advantages. So computer-to-plate (CTP) is a technology avoiding film chemistry, and thereby saving the environment from unnecessary damage. A quicker run-up of the production results in producing less garbage, using less resources, especially using less paper, using less ink, and using less water

8.4.18 Information About CIP3

Detailed information about CIP3 activities - including the latest specification, press material, other related information, the current list of members, and information about membership - can be found on the web at address:

http://www.igd.fhg.de/www/igd-a1/cip3/

9 Outlook: From Printing to Cross Media Publishing

Printing will not go out of business for the foreseeable future. Other media however are pressing into the market of information and communication. Today preparation of material for single media is the rule; this will change over the next few years.

Preparing information for publication is a time-consuming task. This task has not become smaller by the use of electronic media and modern communication. In many cases information is needed for different media, e.g. for an information brochure, for a newspaper, for a glossy journal, for radio and television broadcasting. Preparing information for one of these media is only a little bit less effort than preparing the information in such a way that it can be used by all. Information reuse or multiple use of information is required to keep down overall cost.

One of the key requirements in this respect is the availability of one or more formats that can hold all the required information together. Most formats common today are good in some respect, rarely are they useable in all respects. There are distinct requirements for different media usage.

It is save to state that paper will be around for a while, and quality requirements for reproduction will not become unimportant; still other media will also take up part of the information market. The different products will clearly serve distinct market segments, and will continue to coexist, and to add to the usefulness of information delivery at large.

There are still enough open issues in the field of "Document Imaging", and it is hoped that this book will make a contribution to a better understanding of the problems and solutions.

Bibliography and References

[AD90]   Adobe Systems Incorporated, *PostScript Language Manual – 2nd ed.*, Addison-Wesley 1990

[AD90a]   Adobe Systems Incorporated, *Adobe Type 1 Format*, Adobe Systems Incorporated 1990

[AD90b]   Adobe Systems Incorporated, *Adobe Acrobat Products & Technology – An Overview*, Adobe Systems Incorporated 1990

[AD92]   Adobe Systems Incorporated, *PostScript Language Document Structuring Conventions Specification*, Version 3.0, Adobe Developer Support, 1992

[AD92a]   Adobe Systems Incorporated, *Adobe Type 1 Font Format; Multiple Master Extensions*, Adobe Developer Support, Technical Note #5086, 1992

[AD92c]   Adobe Systems Incorporated, *Encapsulated PostScript File Format Specification*, Adobe Developer Support, Version 3.0, 1992

[AD92d]   Adobe Systems Incorporated, *Adobe Illustrator File Format Specification*, Adobe Developer Support, Draft Version 3.0, 1992

[AD93]   Adobe Systems Inc., *Portable Document Format Reference Manual*, Addison-Wesley, 1993

[AD93a]   Adobe Systems Incorporated, *The Type 42 Font Format Specification*, Adobe Developer Support, Technical Note #5012, 1993

[AD94]   Adobe Systems Incorporated, *Updates to the Portable Document Format Reference Manual*, Adobe Developer Support, Techn. Note #5156, 1994

Bibliography

[AD94a] Adobe Systems Incorporated, *Type 1 Font Format Supplement*, Adobe Developer Support, Technical Note #5015, 1994

[AL92] Aldus Corporation, *TIFF – Tagged Image File Format – Revision 6.0 Final*, Aldus Corp. 1992

[AL93] Aldus Corporation, *OPI – Open Prepress Interface Specification 1.3*, 1993

[AN88] ANSI IT8.1, *User Exchange Format for the Exchange of Color Picture Data between Electronic Prepress Systems via Magnetic Tape*, NPES 1988

[AN88a] ANSI IT8.2, *User Exchange Format for the Exchange of Line Art Data between Electronic Prepress Systems via Magnetic Tape*, NPES 1988

[AN90] ANSI IT8.5, *User Exchange Format for the Exchange of Monochrome Image Data between Electronic Prepress Systems via Magnetic Tape*, NPES 1990

[AN93] ANSI IT8.8, *Graphic Technology, Prepress Digital Data Exchange, Tag Image File Format for Image Technology (TIFF/IT)*, NPES 1993

[BI96] Tim Bienz, Richard Cohn, and James R. Meehan, *Portable Document Format Reference Manual*, Second Edition, Adobe Systems Inc., 1996

[BO86] Boff, K. R., Kaufmann, L., Thomas, J. P., *Handbook of Perception and Human Performance*, vol. 1, New York, Chichester, Brisbane, Toronto, Singapore 1986

[CC85] CCITT, *Standardization of Group 3 facsimile apparatus for document transmission*, Recommendation T.4, Volume VII, Fascicle VII.3, Terminal Equipment and Protocols for Telematic Services, CCITT 1985, pp. 16-31

[CC85a] CCITT, *Standardization of Group 4 facsimile apparatus for document transmission*, Recommendation T.6, Volume VII, Fascicle VII.3, Terminal Equipment and Protocols for Telematic Services, CCITT 1985, pp. 40-48

[CI95] *CIP3 Print Production Format (Version 1.0)*, Fraunhofer Institute for Computer Graphics, Darmstadt 1995

[CI96]   *CIP3 Print Production Format (Version 2.0)*, Fraunhofer Institute for Computer Graphics, Darmstadt 1996

[FI92]   Fink, P., *PostScript Screening: Adobe Accurate Screens*, Adobe Systems Incorporated 1992

[FL75]   Floyd, R. and Steinberg, L., *An Adaptive Algorithm for Spatial Gray Scale,* in Society for Information Display 1975 Symposium Digest of Technical Papers, 1975

[HE95]   Heidelberger Druckmaschinen, *Farbe und Qualität*, Heidelberg 1995

[HE90]   Hell, *ChromaLink*, Dr.-Ing. Rudolf Hell GmbH 1990

[IF95]   *Recommendations for basic requirements of Artwork Transmission System (ATS)*, IFRA Special Report 6.14.4, 1995

[IS86]   ISO/IEC, *Information technology – Text and office systems – Standard Generalized Markup Language (SGML)*, ISO/IEC IS 8879, 1986

[IS92]   ISO/IEC, *Information technology – Digital compression and coding of continuous-tone still images*, ISO/IEC IS 10918, 1992

[IS92a]   ISO/IEC, *Information technology – Computer graphics – Metafile for the storage and transfer of picture description information*, ISO/IEC IS 8632, 1992

[IS93]   ISO/IEC, *Information technology – Computer graphics – Image Processing and Interchange – Functional specification – Part 3: Image Interchange Facility*, ISO/IEC DIS 12087, 1993

[IS93a]   ISO/IEC, *Information technology – Text and office systems – Standard Page Description Language (SPDL)*, ISO/IEC IS 10180, 1993

[IS96]   ISO/IEC, *Information technology – Text and office systems – Document Style Semantics and Specification Language (DSSSL)*, ISO/IEC IS 10179, 1996

[IS97]   ISO 12642, *Graphic Technology, Prepress Digital Data Exchange, Input data for chracterization of 4-colour process printing targets,* 1997

[KN87]   Knuth, D., *Digital Halftones by Dot Diffusion*, in: ACM Transactions on GRaphics, vol. 6, No. 4, October 1987, pp 245-273

KO92]   Kodak, *Kodak Photo CD Products – A Planning Guide for Developers*, Eastman Kodak Company, 1992

[MI90]   Microsoft Corporation, *TrueType Font Files*, 1990,1991,1992

[NE35]   Neugebauer, H. E. J., *Zur Theorie des Mehrfarbenbuchdruckes*, Diss. Dresden 1935

[NE37]   Neugebauer, H. E. J., *Die theoretischen Grundlagen des Mehrfarbenbuchdruckes*, in: Zeitschrift für wissenschaftliche Photographie, Photophysik und Photochemie 1937, Band 36, pp 73-89

[NE37a]   Neugebauer, H. E. J., *Theorie des Vierfarbendrucks mit einem Schwarzdruck*, in: Zeitschrift für wissenschaftliche Photographie, Photophysik und Photochemie 1937, Band 36, pp 169-170

[NE49]   Neugebauer, H. E. J., *Zur Theorie des Mehrfarbenbuchdruckes*, ergänzt durch weiterführende Arbeiten aus der Zeitschrift für wissenschaftliche Photographie, Photophysik und Photochemie 1937-1949, mit einem Vorwort von W. König, (collection of [33,34,35] and some other articles by the author with a preface by W. König, German)

[SA96]   Sandia National Laboratories, *HTML Reference Manual*, http://www.sandia.gov/sci_compute/html_ref.html, 1996

[SC88]   Scitex, *HandShake – Foreign File Transfer Protocol*, Scitex Corporation Ltd. 1988

[UL87]   Ulichney, R., *Digital Halftoning*, MIT Press, Cambridge/Mass. 1987

[WE84]   Terry A. Welch, *A Technique for High Performance Data Compression*, IEEE Computer, vol.17 no.6, June 1984

[WI91]   Wiedling, H.-P. and Daun, S., *EuroRip – A European Development of a Raster Image Processor for Page Description Languages with Common Font Resources*, in: J. Encarnação (ed.), Eurographics'91 Graphics Research and Development in European Community Programmes, pp. 115-127, Eurographics Association 1991

[WY82]   Wyszecki, G, Stiles, W. S., *Color Science – Concepts and Methods, Quantitative Data and Formulae*, 2nd edition, New York, Chichester, Brisbane, Toronto, Singapore 1982

[ZI77]   Ziv, J. and Lempel, A.; *A universal algorithm for sequential data compression;* IEEE Transactions on Information Theory; vol. IT-23; no. 3, pp. 337, May 1977

Glossary

accurate screening

PostScript Level 2 term for closely approximating screening angle and frequency for screens to be used in color printing.

achromatic composition

Achromatic composition removes the achromatic part, i.e. the parts of equal amount of the chromatic colors Cyan, Magenta, and Yellow, completely and substitutes it for Black.

additive color model

Generating a color stimulus by adding light of different color perception to produce a perception of the mixed colors; typically used in display monitors using phosphors generating Red, Green and Blue color perception if singled out, and e. g. producing Yellow by combining Red and Green or producing Magenta by combining Red and Blue.

Advertisement Transmission System

An IFRA recommendation for electronic delivery of advertisements.

AM screen

AM screens or clustered dot halftones or screens have their name from the fact that the tonal value is produced by dots whose size is manipulated, i.e. pixels are clustered to dots of different size in order to obtain

153

different tonal values; screen dots are ordered on a regular (mostly square) grid normally under some (screening) angle relative to the axes.

ANSI IT8

ANSI committee dealing with Graphic Arts (mirror to ISO TC130).

ANSI

American National Standards Institute.

ASN.1

ASN.1, the Abstract Syntax Notation One, is a binary encoding method using type/value pairs; it has a certain distribution in the PTT world; it was used as encoding for the ODA/ODIF family of office standards no longer relevant today; it is also used in the Image Interchange Facility IIF of the ISO standard IPI (Image Processing and Interchange).

ATS

see *Advertisement Transmission System.*

autotypical color model

A combination of additive and subtractive color model, typical for quality printing; the overprinting of different color separations with different inks uses inks as filters that extinguish certain wavelengths of light (subtractive); on the other side the spots of combined ink are so small that they form a partition of space with different light distribution reflected that is integrated by the human viewing system (additive).

brightness

Brightness is the attribute of a visual sensation according to which an area appears to emit more or less light (as defined in CIE International Lighting Vocabulary)

CGM

see *Computer Graphics Metafile.*

ChromaLink

ChromaLink [HE90] is a proprietary exchange format of the former Dr.-Ing. Rudolf Hell GmbH; it allows up to 12 bits per color component for contone images; it also specifies the integration of line work and continuous tone data with line work being supplied at a six times higher resolution than continuous tone.

chromatic color addition

Chromatic color addition is commonly used to support image shadows especially in the neutral grays by adding equal amounts of Cyan, Magenta, and Yellow on top of an achromatic composition; this method leads to a very stable printing process and produces good quality.

chromatic composition

With chromatic composition all color tones are produced using the chromatic process inks Cyan, Magenta, and Yellow. Black is only used to improve reproduction of image shadows and to support contours.

CIE L*a*b*

L*a*b* is derived from XYZ by non-linear transformations to provide an (almost) perceptually equal distance color space.

CIE L*u*v*

L*u*v* is derived from XYZ by non-linear transformations to provide an (almost) perceptually equal distance color space.

CIE lightness

CIE lightness is the L* value of L*a*b* or L*u*v* triples.

CIE luminance

CIE luminance is the Y value of the CIE XYZ 1931 triple.

CIE XYZ

XYZ tristimulus values are based on standardized spectral weighting functions of the cones in the retina; they allow to determine the luminance Y and two other values, called X and Z (related to the three different cones); the magnitude of XYZ components are proportional to physical energy.

CIE Yxy

Y is the same luminance component as in XYZ; x and y are called chromaticity coordinates, dealing with "pure" color in the absence of brightness.

CIE

Commission International de l'Éclairage, International Commission for Lighting.

CIP3

International Cooperation for Integration of Prepress, Press, and Postpress.

CMY

Cyan, Magenta, and Yellow, used as printing colors; based on subtractive color combination to generate color stimuli by absorbing certain ranges of the spectral distribution of the illumination.

CMYK

Cyan, Magenta, Yellow, and Black (key) used as printing colors; based on same model as CMY, but using Black to increase the contrast of images, and to reduce the amount of ink necessary in one spot to produce certain printing colors; neutral gray components of a color (generated by overprinting Cyan, Magenta, and Yellow) are substituted by Black (only needing 1/3 of the ink that would be used with CMY).

color balance

Color tones are reproduced in four color print by specific amounts of the four process colors. Changes in these amounts cause changes in the color tones. To obtain reproducible results and stable color reproduction, the supply of the four process colors have to be kept in balance. Changes in the Black component are regarded as relatively uncritical by the human viewer. Similarly changes of all chromatic color components in the same direction (increase or decrease) is also regarded as not very critical. Hue changes due to changes of chromatic components in different direction however are regarded as very critical. This is visible in neutral gray areas, and therefore color balance is also often called gray balance.

color calibration

Color calibration is the process of transforming color values from input device all the way to output devices thereby retaining the original color perception as closely as possible; in practical terms color calibration is a very complex process, especially when the output is to print; the International Color Consortium (ICC) have specified color profiles that allow to retain the necessary information to the end; correct interpretation can help obtaining a close match to the original.

color gamut

Volume of colors that can be rendered by a certain output device; color gamut is typically shown in CIE xy coordinates in CIE Yxy color space; e.g. for a color display this has typically the shape of a triangle in xy coordinates with the primary colors (of the phosphors) being at the corners of the triangle; for four color printing the shape is typically a hexagon; a 3D presentation of color gamuts is given by color solids, e.g. by the Rösch color solid.

color management

Color management is the name for tools enabling calibrated color to be possible; these tools can be part of the operating system, or can be separate application programs that run under user control; they typically apply ICC profiles to obtain acceptable results.

color rendering dictionary (CRD)

A color rendering dictionary is a PostScript means to provide device independent color to be rendered on a device; it contains the mapping of a CIE based color space to the device color space.

color temperature

Spectral power distribution radiated from a hot object – according to Planck called a black body radiator – is a function of the temperature to which the object is heated; many sources of illumination have, at their core, a heated object, so it is often useful to characterize an illuminant by specifying the temperature (in units of kelvin, K) of a black body radiator that appears to have the same hue.

Computer Graphics Metafile

An ISO standard for exchanging pictures generated by computer graphics.

computer-to-plate (CTP)

CTP is a technology that directly generates the plate from the digital image generated by the RIP; it avoids the step of producing film, and thus also avoids film chemistry harmful to the environment.

cones

Three different types of sensors, responsive to different ranges of wavelengths of the visible portion of light, located in the retina. The specific wavelength sensitivity of the different types of cones is the foundation of color vision. The absorption spectra of the three types of cones have their maxima roughly at 445 nm, 535 nm, and 600 nm.

control elements

Objects used for controlling certain parameters in the printing process chain; examples are color control strips made up of full tone elements (to control ink coating thickness), raster elements (to control dot gain), color balance elements (to control the balance between the different printing inks), and slur and doubling elements (to control behavior of the rubber sheet on offset presses).

CRD

see *color rendering dictionary*.

crossover

a publication object spanning two pages side by side, e.g. a heading or an image; of course these should align well across the two pages; therefore crossings are often placed in the centerfold of a publication

CTP

see *computer-to-plate*.

damping unit

The damping unit applies water to the surface of the plate via the damping duct and vibrator rollers, the damping distributor and the damper roller.

Delta E

Color difference measurement that is built on perceptually equal distant color coordinates, typically L*u*v* or L*a*b*; in most cases L*a*b* is used; otherwise the color coordinate system is specified in an index. Delta E is calculated as the euclidian distance in the color coordinate system chosen.

display list

Display lists are representations of the content of a page that allow fast rendering to device pixels; there are various forms of display lists in use, different forms exhibiting different strengths and weaknesses; they may already be device dependent (in most cases), or they may be device independent; they may contain non-self-overlapping areas as primitives, or they may contain only objects that no longer overlap other objects; the most time consuming step is performed out of the display list, the rasterization, or screening.

159

document structuring conventions (DSC)

DSC is a method of structuring a PostScript file in a way that allows to detect important properties (e.g. where do pages start, how many pages are in this file, which fonts are used in this file) without interpreting the file; it also is a convention that allows to make pages in the file independent of each other, which is good programming practice, but a property not enforced by mere PostScript.

Document Style Semantics and Specification Language (DSSSL)

DSSSL is a standard that allows to map tags in SGML to a concrete typographic presentation.

Document Type Definition (DTD)

A DTD is the definition of a legal document in SGML.

dot gain

During printing, due to different effects, an increase of the area carrying ink can be observed. This effect is called dot gain, because it results in an increase of the dot size of AM screen dots; its main effect is the spread of ink from the ink carrying region to the regions normally not carrying ink; the effect will be more noticeable if there are longer borderlines between ink carrying and no ink carrying regions; this model exhibits small dot gain in low and high tonal values, and increased dot gain in the middle tonal range, and it is much more noticeable in FM screens, than in AM screens, because there are more unconnected pixels, and hence longer borderlines. Another definition of dot gain is the difference between tonal values on the film and in print.

dot join

Dot join is the increase of the inked area due to ink flowing together under pressure at positions where screen dots connect. Dot join is more visible, if screen dots connect in both axis directions at the same tonal value; to avoid this effect, elliptic dot shapes are used.

doubling

Development of lighter duplicates of objects in offset printing shifted from the original slightly, often due to problems with the rubber sheet.

DSC

see *Document Structuring Conventions*.

DSSSL

see *Document Style Semantics and Specification Language*.

DTD

see *Document Type Definition*.

Encapsulated PostScript (EPS)

An EPS file contains only one single picture; it starts out with simple comments that give the originator, the resources used (e.g. fonts), the bounding box, and other useful information, in a format that can easily be scanned by application software without interpreting PostScript; operators that induce global and irreversible changes to the PostScript interpreter's internal state are not allowed in an EPS file; it may contain a low resolution raster image of the content of the file, which the including program may be able to display; the main intention of EPS is to allow integration of different PostScript objects coming from different applications.

EPS

see *Encapsulated PostScript*.

eye response functions

Functions of wavelength, specifying the response of cones to light of different wavelength; these functions $\bar{x}$, $\bar{y}$, and $\bar{z}$ have been standardized by CIE according to their standard 2° observer (for small areas of color) and 10° observer (for large areas of color).

hinting

Hinting is a process associated with fonts that ensures that certain properties of fonts are retained, especially in small font sizes (in number of pixels); e.g. it is necessary to make special provisions for serifs not to disappear at sizes smaller than 10 pixels.

HPGL

HPGL is the name of the proprietary Hewlett-Packard graphics language.

HSB

see *hue-saturation-brightness*.

HTML

see *Hyper Text Markup Language*.

hue

Hue is the attribute of visual sensation according to which an area appears to be similar to one of the perceived colors, Red, Yellow, Green and Blue, or a combination of two of them (as defined in CIE International Lighting Vocabulary).

hue-saturation-brightness (HSB)

Color model that is flawed with regard to the human color vision; it is used in PostScript as an alternative to specifying color in RGB; if predictable color is intended, a CIE based color system should be used rather than HSB.

Huffman encoding

Huffman encoding is an encoding method that assigns shorter codes to more frequently occuring items, and longer codes to more seldom occuring items; Huffman encoding is applied for example in facsimile encoding.

hydrophilic

Attracting water or friendly to water.

hydrophobic
> Not attracting water.

Hyper Text Markup Language (HTML)
> HTML is the specific SGML document type definition that governs documents in the World Wide Web.

ICC
> see *International Color Consortium.*

IFRA
> INCA-FIEJ Research Association, the research institute of newspaper publishers.

IIF
> Image Interchange Facility, part of the ISO standard on Image Processing and Interchange.

image interpolation
> Interpolation used for interpretation of an image assumes the color values in a cell of the image array is correct for the center of the cell; color values that map to a position different from the center of the cell are evaluated by a bilinear interpolation of the participating four color values of the neighboring pixels.

image sampling
> Sampling used for interpretation of an image determines the color value of a device pixel by inverse mapping of the device coordinates of the pixel to the image array. If the image array cell is large compared to the device resolution (which is normally the case) this leads to the effect that the pixels in the area of the cell all are set to the same color value (i.e. the value of the cell).

image setter
> Image setters are high precision devices that produce device pixels on a light sensitive material applying a laser beam directing the light to the spot to be imaged through an arrangement of polygonal mirrors. Depend-

165

ing on the material to be imaged (different kinds of film, plates), laser light of different wavelengths is used. After the imaging process normally a development process is necessary to fix and stabilize the image on the material.

imposition

Imposition is the process that maps pages to sheets which later shall be printed (e.g. on an offset press); the result of placing and orienting pages on a sheet is called a signature.

impression cylinder

The ink image on the rubber sheet is applied to the paper that is fed with pressure between the rubber sheet cylinder and the impression cylinder.

ink keys

The amount of ink used for printing is controlled across the width of the press; the ink keys control the amount of ink supplied in each of several ink zones across the press.

ink trapping

Ink trapping is the effect that ink adhesion is better on paper than on top of an already printed ink, especially if printing wet-on-wet; this results in less ink in an area than expected from the area coverage defined for that ink in this area.

inking unit

The inking unit is that part of the press that transfers ink to the plate; ink flow starts with the ink duct (which contains the ink supply); ink is transferred via the ink duct roller and the ink vibrator roller (to control the amount of ink supplied); further rollers are used to reduce the thickness of the applied ink coating to the required measure of roughly one micron; the plate inkers finally transfer the ink to the plate.

International Color Consortium (ICC)

Group that has specified so-called device profiles that allow to define the exact relation of colors on one device (e.g. a scanner) to colors on another device (e.g. a printer); this is achieved by going via a CIE based intermediate color system; rendering intents in such profiles allow to also specify the behavior of devices in cases when colors do not fall within the device's gamut, and also allow to indicate the relative importance of color properties.

irrational screen

An irrational screen approximates the correct angles with an error that is small enough that unwanted visual effects (moiré) disappear or are at least minimized. This can be done by different methods. One method is to sample an accurate function over space to generate the proper distribution of the screen dots and thereby approximate the angle more correctly. Another way of achieving more accurate approximation of angles and still use repeatable patterns to generate the screen is the technique of super-cells. Instead of a defining single cell and repeating it all over the page, a super-cell containing many screen cells is applied, and thereby approximating the angle more exactly. The super-cells are computed according to the first method.

ISO TC130

Technical Committee 130, ISO committee dealing with Graphic Arts Standardization.

ISO

International Organisation for Standardisation.

ISO/IEC JTC1

see *JTC1*.

ISO/IEC JTC1/SC24

see *SC24*.

JPEG

ISO standard for image compression; it contains both methods for lossy as well as lossless compression; in most cases JPEG is used synonymous with JPEG base–line, which is a lossy compression method employing discrete cosine transformation (DCT) to cut off high frequency data, thereby omitting details that are not important in some applications.

JTC1

ISO/IEC JTC1 is a Joint Technical Committee of ISO and IEC dealing with information technology.

kerning

Adaptation of glyph distances in words to give a regular visual impression.

Lempel-Ziv (LZ77)

LZ77 is a lossless compression method, the predecessor of the LZW compression.

Lempel-Ziv-Welch (LZW)

LZW is a lossless compression method named after their inventors; it is well suited for text and line art images; it does not well compress photographic images.

light gathering

Light entering paper and being dispersed in paper and trapped under neighboring areas covered with ink; this leads to darker image perception than would be expected from the ink coverage.

light trap

see *light gathering*.

lightness

Lightness is the perceptual reponse to luminance; lightness perception is roughly logarithmic; see also CIE lightness

luminance

Luminance is radiant power weighted by a spectral sensitivity function (see also CIE luminance).

LZ77

see *Lempel-Ziv*.

LZW

see *Lempel-Ziv-Welch*.

Mac Adams ellipses

CIE XYZ (and Yxy) has bad properties regarding perceived visual distance; the distances of colors that are just distinguishable by the human visual system are very different in different areas of the color space; these differences are illustrated in form of so-called Mac Adams ellipses (in 3D diagrams ellipsoids) showing the distance of colors that are just distinguishable for the human viewer.

metamere colors

Object colors that are perceived as the same colors under certain illumination, although the objects have completely different remission curves.

metameric index

The metameric index is the perceptional color difference of two objects perceived as identical color under one illuminant, if viewed under different illumination conditions.

multiple master font

Multiple master font is a font technology of Adobe that allows to represent a font not available by another font of similar properties.

Neugebauer equations

Equations named after Hans E. J. Neugebauer, one of the pioneers in the area of color reproduction, specifying the interaction of areas of different printing inks and paper (white), adding up the area fractions multiplied with the respective CIE XYZ values of the respective area; the model assumes certain statistical relationships between ink covered areas according to their relative area coverage. Originally based on 3 color printing, it has been extended to also cover 4 and more color reproduction.

oleophilic

Friendly to oil (or ink).

oleophobic

Not attracting oil or ink.

Open Prepress Interface (OPI)

OPI was introduced by Aldus Corporation as a set of PostScript comments that are used together with low resolution copies of originally high resolution images; the comments provide information about the location of the related high resolution files together with placing, cropping, scaling, rotation, and color information; since this information is syntactically hidden in PostScript comments, it can be processed without interpretation of the PostScript code.

OPI server

OPI servers are programs that calculate low resolution samples at scanning, and substitute the low resoution version with the high resolution image before sending a PostScript job to the RIP

OPI

see *Open Prepress Interface.*

opponent color theory

Theory basing color perception on synaptic combinations of basic color responses of the three types of cones, thus forming opponent pairs Red-Green, Yellow-Blue, and dark-bright.

PackBits encoding

see *runlenght encoding.*

page description language (PDL)

A PDL is a (programming) language allowing to describe output on pages in a device independent way; examples of PDLs are PostScript, and its predecessor, Interpress.

path

A path is the geometric description of the outline of an area; it may contain curves and lines as boundary elements; a path may contain multiple disjoint sub-paths, and the described areas may be overlapping; paths can be used to be stroked, filled, or used as a clip area; two different inside rules can be applied, the even-odd and non-zero-winding-number rule.

PCL5

Proprietary Hewlett-Packard page description language.

PDF

see *Portable Document Format*.

PDL

see *page description language*.

perceptually equal distance color systems

Color systems that inherently have a distance metric close to perceived distance for a human viewer; typical representatives of such systems are L*u*v* and L*a*b* color coordinates.

PhotoCD

PhotoCD is a proprietary image format of Kodak on compact disc; it stores images at multiple fixed resolutions, uses a device independent color system, and a proprietary compression method which is not made publicly available.

plate copier

Machine that maps images on film to the plate using an analog copying process.

plate cylinder

Cylinder in the printing press on which the plate is mounted.

plate setter

Image setter generating output on printing plates.

Portable Document Format (PDF)

PDF is essentially a "flat" PostScript, which means that no procedures and control constructs are supported; it is the format that is employed for Adobe's Acrobat family of products; PDF contains over the mere page description also features to be used in interactive and networking environments (e.g. WWW).

PostScript Level 1

PostScript Level 1 was introduced only after a Level 2 PostScript had been described; it served as the label for the original PostScript as brought out in the mid-eighties; in addition it included some color extensions, which were introduced, because prepress applications could not use PostScript without such extensions; part of these extensions were operators supporting the CMYK color space.

PostScript Level 2

PostScript Level 2 provides over Level 1 patterned fill for graphics; it also provides improvements in areas the user does not normally recognize, like better and controllable free storage handling in interpreters, it provides additional simple graphics primitives, it provides a more efficient handling of text, and, most important, it offers a complete color concept with device dependent and device independent color.

PostScript Level 3

Latest announcement of Adobe regarding PostScript; Level 3 is announced to be a combination of PostScript Level 2 and PDF.

PostScript

PostScript is a page description language (PDL). It can be characterized as stack oriented pogramming language; it uses postfix notation of operators, i.e. operands precede the operators; the language is processed sequentially on a token by token basis; PostScript is not only a programming language, it also provides mechanisms for rendering, for placing marks on a page; a PostScript file is a program normally encoded in ASCII describing the graphical content of a page in a device independent way.

PPF

see *Print Production Format.*

preflight systems

Systems that perform certain plausibility checks on the input files to find missing or wrong information before costly processing happens further down stream; often including a PostScript RIP to check processibility of the PostScript code; however this solution does not catch 100% of the problems.

print characteristic

The tonal value with dot gain (in print) as a function of the tonal value in film.

Print Production Format (PPF)

File format specified by CIP3 to support computer-integrated manufacturing of print products.

PTT

PTT is an acronym for all post and telephone/telegraphy companies world wide; today this means providers for telecommunication services.

raster image processor (RIP)

RIP is used for the processor that converts PostScript, or more general, a PDL to the device's representation – the output is directly usable for the destination device – or for the processor that converts a continuous tone

173

signature

Result of placing and orienting pages on a sheet, typically done by imposition software or manually.

simultaneous contrast

Identical color patches are perceived as different colors if surrounded by different environment.

slur

Expansion of points either in print direction or sideways.

SPDL

see *Standard Page Description Language.*

spot color

Specific colored ink used in excess of the process colors C, M, Y, and K for reproduction of colors otherwise not well reproducible; also necessary for reproduction of special effects (silver, gold, metallic).

spectral remission

The amount of light of a certain wavelength that is remitted from a surface.

Standard Generalized Markup Language (SGML)

The Standard Generalized Markup Language (SGML) is a language to specify a document markup syntax; with SGML the legal logical structure of a document is defined in a so-called document type definition (DTD), and with appropriate editors you can ensure that only correct documents according to the given DTD are produced; HTML, the markup used for the World Wide Web, is basically a SGML DTD.

Standard Page Description Language (SPDL)

SPDL is an ISO standard that provides basically the semantics of PostScript Level 2 with different syntax (using ASN.1).

subtractive color model

The overprinting of different color separations with different inks uses inks as filters that extinguish certain wavelengths of light (subtractive) and thus form a color perception; this model is also used if mixing different dyes, e.g. mixing Cyan and Yellow dyes to produce Green.

Tagged Image File Format (TIFF)

TIFF is a very common format for raster images; as a tagged format it provides a set of tags to supply attributes and image data values as tag-value pairs; TIFF is a binary format that allows to describe bi-level, grayscale, palette-color (indexed color), and full color images in several color spaces; TIFF allows for different ordering or interleaving schemes of the image data; It also allows for segmenting images into small rectangular subregions – so-called tiles – in order to improve accessibility in different orders.

TIFF

see *Tagged Image File Format.*

tile

A tile is a rectangular (or square) subregion of an image; it allows to structure large images into a lattice of tiles thereby enabling improved access strategies to the large volume of data.

trapping

Overprinting multiple separations requires exact alignment of the different separations to avoid unpleasant effects. For example, if one places a Black text on top of a Cyan area, a slight misalignment of the two involved separations (C and K) results in white shadow regions around the text. Trapping is a method that allows to relax the strict requirement on sub pixel alignment. In our example, the Cyan separation would extend under the text glyphs; the size of extension that is necessary depends on the process to be used for printing. Trapping does not eliminate completely all

177

effects of misalignment, but it reduces the visibility of these effects significantly.

tristimulus color theory

Theory basing color perception on the three different types of cones and their response to light; this theory also forms the basis for the popular RGB device color system.

UCR

see *under-color removal*.

under-color removal (UCR)

Under-color removal or under-color reduction is a technique that uses chromatic composition, but a part of the three overprinted chromatic inks is substituted by Black ink.

white point

Specifies ideal pure white; the white point is dependent upon the illumination, i.e. the white points for different standard illuminants differ; the human eye adapts color perception to the white point, and perceives colors relative to the white point

World Wide Web (WWW)

WWW is the interactive access to all services (e.g. electronic mail, file transfer, information services) of the Internet via a uniform user interface; it allows direct access to information resources distributed throughout the world; the interactive tools used to access the WWW are called browsers; WWW, or the web, is also used to signify the whole network that is accessible through web browsers.

WWW

see *World Wide Web*.

Index

Symbols

A

Springer
and the
environment

At Springer we firmly believe that an international science publisher has a special obligation to the environment, and our corporate policies consistently reflect this conviction.

We also expect our business partners – paper mills, printers, packaging manufacturers, etc. – to commit themselves to using materials and production processes that do not harm the environment. The paper in this book is made from low- or no-chlorine pulp and is acid free, in conformance with international standards for paper permanency.

Springer

Printing and binding: Druckerei Triltsch, Würzburg